SIZING AND SELECTING AIR POLLUTION CONTROL SYSTEMS

HOW TO ORDER THIS BOOK

BY PHONE: 800-233-9936 or 717-291-5609, 8AM–5PM Eastern Time

BY FAX: 717-295-4538

BY MAIL: Order Department
Technomic Publishing Company, Inc.
851 New Holland Avenue, Box 3535
Lancaster, PA 17604, U.S.A.

BY CREDIT CARD: American Express, VISA, MasterCard

SIZING AND SELECTING AIR POLLUTION CONTROL SYSTEMS

Edited by
HOWARD E. HESKETH, Ph.D., P.E.
FRANK L. CROSS, JR., P.E.

LANCASTER · BASEL

Sizing and Selecting Air Pollution Control Systems
a **TECHNOMIC®** publication

Published in the Western Hemisphere by
Technomic Publishing Company, Inc.
851 New Holland Avenue, Box 3535
Lancaster, Pennsylvania 17604 U.S.A.

Distributed in the Rest of the World by
Technomic Publishing AG
Missionsstrasse 44
CH-4055 Basel, Switzerland

Printed in the United States of America
10 9 8 7 6 5 4 3 2 1

Main entry under title:
Sizing and Selecting Air Pollution Control Systems

A Technomic Publishing Company book
Bibliography: p.
Includes index p. 161

Library of Congress Catalog Card No. 93-61892
ISBN No. 1-56676-126-3

Table of Contents

Preface

NEW standards are being required to meet best available control technology (BACT) and maximum achievable control technology (MACT). These new standards require high-efficiency collectors, and in many instances they require test data to determine the appropriate sizing of the control system.

Some of the information that should be available for the design and operation of an air pollution control system includes:

- production rate
- fuel and raw material composition and rate
- system gas flow rate, calibrated flow measurement equipment and inducted draft (ID) fan power input
- system resistance (pressure drop), including ID fan static pressure, collector resistance and duct and stack resistances
- stack gas characteristics including temperature, gas composition, moisture content, particle composition and size, size distribution and overall corrosive potential
- control equipment function indicators
- ancillary equipment function indicators
- process operating procedures (process changes can often end in different test results, which makes it difficult to specify equipment)

This book is designed to acquaint the reader with current regulations and with the necessary information to size air pollution control systems. The material presented should also help enable one to select the appropriate equipment for retrofit or new process control, to prepare specifications to purchase equipment, and to prepare permits for air pollution control systems.

The editors believe that this book will provide guidance to help those responsible for air pollution control to specify systems which are cost-effective and energy-efficient to meet the needs of their employers and the government. When equipment specifications are properly prepared, they provide for an easier comparison of competitive bids of those devices capable of meeting standards reliably and economically.

FRANK L. CROSS, JR.
HOWARD E. HESKETH

Acknowledgements

THE editors wish to thank CTA for the initial concept of this book. We greatly appreciate the work of Amy Dunn for preparing the manuscript. Special thanks also go to Judi Cockrum and the Southern Illinois University Department of Mechanical Engineering and Energy Processes for facilities and help. Finally, we acknowledge the assistance of Scott Lane, engineering graduate student, who helped review the manuscript, and to the SIU Fall 1992 Air Pollution Control class who reviewed the galleys.

CHAPTER 1

Equipment Sizing Data

HOWARD E. HESKETH, Ph.D., P.E.[1]

1.1 INTRODUCTION

AIR pollution control (APC) systems must perform to meet the needs of the process from which emissions are to be controlled and to satisfy the applicable emission regulations. Acquiring APC systems is much like buying and operating a vehicle. For example, many types of cars are available to transport a person from point A to point B. There are different capital costs associated with the various models and also different operating costs. Likewise, the convenience and output response differ.

Extending this example, a truck may be required instead of a car, so a whole new group of decision-making parameters are introduced. APC equipment is similar and will be discussed in detail in later chapters of this book. This section is intended to provide basic design input needs, so models, types and sizes of APC equipment can be specified intelligently.

1.2 REGULATIONS

Regulations govern what is permitted to be released from the APC equipment. These will be covered in detail later in this book. However, a short introduction and explanation of a few basic terms and concepts are given here.

Historically, the Clean Air Act Amendments (CAAA) of 1977 specified control levels of BACT (best available control technology) and LAER

[1]Southern Illinois University at Carbondale, Carbondale, Illinois.

(lowest achievable emission rate). BACT is technology based and used for new source review to prevent significant deterioration (PSD). LAER was required in non-attainment areas. RACT (reasonably available control technology) was specified by the CAAA of 1977, but was directed mainly to existing VOC (volatile organic compound) sources in ozone non-attainment areas.

MACT (maximum achievable control technology) is specified by the CAAA of 1990. It is more restrictive and is based on emissions. It is especially directed to all major air toxic emissions. A MACT emission standard is a technology standard defined as the maximum degree of reduction in emissions. Taken into consideration is the cost of achieving such emissions reduction, and any non-air quality health and environmental impacts and energy requirements.

The EPA (Environmental Protection Agency) is to engage in an MACT review for each Air Toxic. In setting MACT, the EPA is also to consider any combination of:

- process changes or substitution of materials
- enclosure of systems or processes
- capture and treatment of fugitive, process, stack and storage emissions
- design, equipment, work practice, or operational standards

A summary of factors considered when setting MACT emission standards is given in Table 1.1 and suggested possible APC equipment needed to meet these emission requirements is noted in Table 1.2.

In order to meet more stringent NAAQS (National Ambient Air Quality Standards) ozone controls, VOC emission limits will be reduced in many states for existing sources. The result will be an increase in control on "major" sources of VOCs, which can include facilities releasing as little as 10 tons VOC/yr. The new MACT standards may require installation of

TABLE 1.1. MACT Control Requirements.

Source Type	Control Standards
New source	Best controlled similar source
Existing source	Average of best 12% of existing sources or LAER if 30 or more sources in category are meeting LAER or Average of best five (5) sources if fewer than 30 sources in category
Area source	Generally available control technologies or management practices

TABLE 1.2. Probable APC Equipment and Procedures Required to Meet MACT.

Air Toxic	Control Concept	Comments
Particulate	Fabric filters	Removal efficiencies ~99%
	High efficiency wet scrubbers	Removal efficiencies >97%
Acid gases	Spray dryer/fabric filter	May be required in combination with thermal oxidation
	Venturi/packed tower	May be required in combination with thermal oxidation
Organic Vapors		
Volatile	Direct flame thermal oxidizer	Most applicable High operating cost
	Catalytic thermal oxidizer	Selective/clean gas stream Low moisture
	Carbon adsorption	Selective/clean gas stream Low moisture
Semi Volatile	Direct flame thermal oxidizer	Most applicable
	West scrubber	Extremely selective

additional control equipment process modifications, regular monitoring or testing, personnel training programs, and developing O & M (operating and maintenance) programs.

1.3 INFORMATION NEEDED TO SIZE APC EQUIPMENT (APCE)

This section is intended to provide information needed to size APCE. It is not intended to tell you how to perform the actual sizing deliberations.

1.3.1 GENERAL

In addition to the emission limits noted in the previous sections, operational data and source uncontrolled emission data are needed, along with the physical description of the type of equipment to be sized and a thorough concept of the entire system of concern.

Materials to be controlled may be either gaseous and/or particulate in nature. Vapors are considered as gases with the notation that they are considered to be easily condensed gases and may thus become particulate matter.

Particles basically have a log-normal (i.e. geometric) size distribution. Therefore, they can be plotted as straight lines on log-probability graphs, as shown in Figure 1.1. Most commonly, the data are by mass. The slope

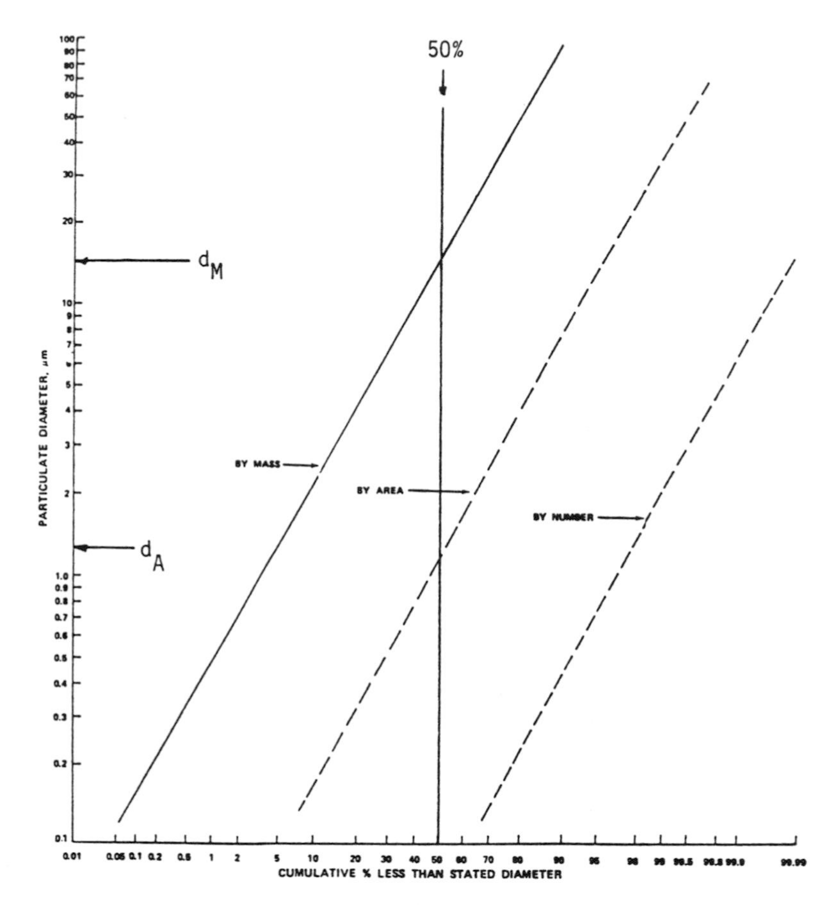

Figure 1.1 Typical log probability cumulative distribution of coal combustion emissions leaving a low efficiency electrostatic precipitator.

of the curve represents the standard geometric deviation, σ_g. The size particle that occurs at the 50% cumulative probability is the mean diameter. For the mass curve, this is indicated on Figure 1.1 as d_M and is the mass mean diameter. One can then proceed to find σ_g by either:

$$\sigma_g = \frac{d_M}{d_{16}}$$

or (1.1)

$$\sigma_g = \frac{d_{84}}{d_M}$$

where d_{84} and d_{16} represent the size particle that occurs at the 84 and 16 cumulative % < by mass, respectively. Note that the σ_g for the area and the number plots would be the same for this same sample. Most commonly, the magnitude of σ_g for airborne particulates is about 1.5−2.5.

Basic information needed to design the APCE is shown in Table 1.3. Units noted are mostly English where:

ppm = parts per million by volume for gases
ppm = parts per million by mass for liquids or solids
acfm = actual cubic feet per minute
scfm = standard cubic feet per minute ("standard" indicates conditions of 68°F and 1 atmosphere)
dscfm = dry standard cubic feet per minute
gr = grains (7,000 gr = 1 pound)
MSDS = Material Safety Data Sheets

TABLE 1.3. General Information for the Design of Air Pollution Control Systems.

Inlet gas stream
Composition/concentration of gases/vapors, % or ppm
Water, %
Temperature
Volumetric flow rate, acfm or scfm or dscfm
Particles present? d_M, σ_g, concentration (gr/dscf)
Dew point
Toxics
Other
Outlet gas
Composition/concentration
Water
Temperature
Flow rate
Saturation
Other
Dust resistivity
Dust cake porosity
Heating value
Volatility/flash point
Mix composition
Toxics
MSDS
Gas flow (ducting) configuration
Space available
Ambient weather conditions
Component pressure drop

The more that is known relative to inlet-outlet conditions, the more exact the APCE design process can be. The MSDS are excellent sources of physical and chemical date.

Ambient weather data are important. Such data can determine how non-adiabatic a component is and can affect condensation, dew points, and freezing of slurries. Bridging in hoppers, "mudding" of filter bags, and corrosion are other negative effects of cold weather.

Pressure drop (ΔP) is critical for almost every type of APCE. It is "easy" to obtain, but it must be done correctly. Try to obtain these static pressure values at points in the duct where gas velocities are similar. Sample at high points and not where they will be influenced by kinetic pressure (e.g., not in an elbow). Keep the sample lines dry and the lines and taps clear. The physical arrangement of inlet and outlet ducting is critical to the operation of many types of APCE. Both gas and particle distribution are affected by this, as well as pressure readings.

1.3.2 SCRUBBERS

Scrubbers can be either wet or dry, as noted in Table 1.4. The wet type must always have mist eliminators (i.e., demisters) and the dry types are followed by a particle collection system. Mist eliminator gas velocity must not be too fast (or reentrainment will occur) or too slow (they will not demist and entrainment will occur).

Scrubbing liquid must be adequate and must be available at the correct position in the wet systems. As a rule-of-thumb, a minimum of about 5 gpm of liquid is needed per 1,000 acfm of gas to be cleaned for particle collection and about 20 gpm/1,000 acfm for gas adsorption in a wet scrubber. Sometimes reactions take place. This requires time. Precipitates may also be formed; and ideally, they should come out of solution on solids present in the recycle liquid, rather than on the internals of the scrubber.

When working with hot, unsaturated gases, a quench should be considered. This can: result in positive phoretic forces; reduce the collector size; protect the APCE from thermal, chemical corrosion and physical erosion; and help wet the particles.

1.3.3 ELECTROSTATIC PRECIPITATORS (ESPs)

ESPs are big boxes which become more efficient as more stages are added in series. Each stage removes about 80% of the stage inlet dust on a well-designed unit. Therefore, the inlet sections require much more frequent cleaning than the outlet stages. This is good, because our major source

TABLE 1.4. **Information Needed for Design of Scrubber (Wet or Dry) Systems.**

- Type of Scrubber
  - Material of construction
  - Size
  - Co-current, counter-current, or cross-current flow
  - Packing—type
  - Packing—size
  - Number of actual stages
  - Superficial gas velocity, ft/sec
  - ΔP
  - Distribution (channeling)
  - Equilibrium diagram
  - Henry's Law Constants
  - Stage efficiency
  - Gas contact time
  - Velocities
  - Other
- Entrainment Separators
  - Type
  - Horizontal or vertical gas flow
  - Velocity, ft/sec
  - ΔP
  - Other
- Scrubbing Liquid
  - Type
  - Recycle gpm
  - Delay/reaction time
  - Solids present? concentration, dissolved/suspended
  - Make-up gpm
  - Bleed gpm
  - Temperature
  - Disposal—toxics
- Quench
  - Time
  - Fresh water gpm
  - Distribution—gas and liquid

of emissions from an ESP comes when the stage is cleaned. As noted in Table 1.5, type and frequency of rapping/cleaning are very important. Dust reactivity is also important and is noted in Section 1.3.1. This affects wire type, cleaning, field strength and gas velocity. The velocity must not be too high, or it will blow collected dust from the collection surfaces. Too low a velocity results in an even larger ESP.

Typical ESPs can achieve up to 99.9% overall collection efficiency with proper design, operation and at an aspect ratio 1.5 (length to height dimensions).

TABLE 1.5. Information Needed for Design of Electrostatic Precipitator Systems.

Material of construction
Rapping/cleaning interval (frequency), type
Corona power
Sparking rate
Wire type
Discharge method
Insulation
Control type
Aspect ratio
Gas velocity

1.3.4 FILTERS

Information needed relative to specifying filters is listed in Table 1.6. Filters are the most efficient particle capture system and good units can operate at efficiencies of up to 99.99%. They must be protected from dew points (both acid and moisture). For highly deliquescent salts, the gas temperature should be at least 80°F above saturation temperature. For other salts, the temperature should be at least 30°F above saturation. Cleaning must be completed frequently to keep pressure drops from becoming excessive.

Another important parameter relative to filter pressure drop is cake porosity. This is noted in Section 1.3.1 (Table 1.3). In addition to keeping the system dry, the dust to be filtered should consist of "large," as well as "small" particles to produce a more porous filter cake.

1.3.5 INCINERATORS

A different group of information data is required for incinerators, as shown in Table 1.7. Incinerators are not actually APCE per se; however,

TABLE 1.6. Information Needed for Design of Filter Systems.

Material of construction
A/C
Coth type
Filter type and configuration
Discharge method
Insulation
Cleaning cycle time
Method for disposal of cake
Toxics
Control type

TABLE 1.7. **Information Needed for Design of Incinerator Systems.**

Type
Material of construction
Refractory
Burners
Feed/discharge methods
Draft
Temperature
Auxiliary fuel
Control type
Toxics
Heat release—area/volumetric

operation procedures greatly influence emission type and amount. This includes emissions such as NO_x, HCl, VOC, PICs (products of incomplete combustion) and dioxins/furans. Temperatures must be high enough and times long enough to destroy undesirable emissions while not creating others.

CHAPTER 2

Permit Requirements for New and Retrofit Systems (Including Toxic Modeling)

JOSEPH L. TESSITORE, P.E. [1]

2.1 CONTROL OF AIR TOXICS

2.1.1 PAST AIR TOXICS CONTROL

BEFORE the 1990 Clean Air Act, air toxic control procedures included NESHAPs, e.g., for benzene and vinyl chloride (some NESHAP vinyl chloride sources and standards are listed in Table 2.1); VOC control for ozone; and state agency approaches—TLV levels (e.g., 1/100 of TLV), acceptable ambient concentration levels (AACL), risk assessment (e.g., 1 cancer case per million).

2.1.2 PAST AND PRESENT

The following definitions serve as a bridge to show how control technologies have progressed:

(1) BACT—"Best Available Control Technology"
- Clean Air Act 1977
- technology based
- new source review for PSD
- applies mainly to add-on controls

(2) LAER—"Lowest Achievable Emission Rate"
- Clean Air Act 1977
- technology and emission based
- new source review in non-attainment areas

[1]Vice President, Cross/Tessitore & Associates, P.A., Orlando, Florida.

TABLE 2.1. National Emission Standards for Hazardous Air Pollutants (NESHAP).

Source	Standard
Subpart F vinyl chloride	
Ethylene dichloride plants	
All exhaust gases	10 ppm (proposed: 5 ppm)
Vinyl chloride plants	
Formation and purification exhaust gases	10 ppm (proposed: 5 ppm)
Relief valves	No discharge
Fugitive emissions	Minimized
Polyvinyl chloride plants	
Exhaust gases from reactor or stripper or mixing, weighing, or holding containers, or monomer recovery system	10 ppm (proposed: 5 ppm)

(3) RACT—"Reasonably Available Control Technology"
- Clean Air Act 1977
- technology and emission based
- mainly for existing VOC sources in ozone non-attainment areas

(4) MACT—"Maximum Achievable Control Technology"
- Clean Air Act 1990
- mainly emission based
- all "major sources" of HAPs
- includes:
  - process changes
  - enclosure or elimination
  - collection and capture (process, storage, stack and fugitives)

(5) GACT MP—"Generally Available Control Technologies or Management Practice"
- Clean Air Act 1990
- all area sources of HAPs

States have the right under the 1990 CAA to establish regulations which are more stringent than federal control regulations. Therefore, in a "tongue-in-cheek" manner, we could note that a "BIG MACT" is a state emission or control standard that is beyond EPA MACT standards.

2.1.3 CLEAN AIR ACT AMENDMENTS OF 1990

The Clean Air Act (CAA) Amendments of 1990 promulgated in 40 CFR:

- Defined "Source," "Modification," "Bubble," and "Owner/Operator"

- Defined HAPs (Hazardous Air Pollutants) and the Listing Process
- Gave Source Categories
- Established MACT (Maximum Achievable Control Technology) Standards
- Established Schedule for EPA to Promulgate MACT Standards
- Gave Residual Risk Standard
- Gave Early Reduction Credits
- Prohibitions and Establishment of Compliance Schedule for New and Existing Sources of HAPs
- State Program Delegations
- Prevention of Accidental Release of HAPs and Other Hazardous Substances

Some of these items covered in the air toxics program (Title III) of these amendments are discussed in detail below.

The definition of a ''major source'' and ''area source'' is given in Section 112(a) as:

1. A major source is ''any stationary source or group of stationary sources located within a contiguous area and under common control that emits or has the potential to emit considering controls, in the aggregate, 10 tons per year or more of any hazardous air pollutant or 25 tons per year or more of any combination of hazardous air pollutants.''
   (a) EPA may regulate a source with a lesser emission rate on the basis of the potency, persistence, potential for bioaccumulation and other factors.
   (b) Radionuclides may be regulated under different criteria.
2. Any stationary source which is not a ''major source'' of HAPs is an ''area source.''
   (a) Motor vehicles are not subject to regulation under this Title and are not area sources.
   (b) Section 112(n)(4) forbids the aggregation of emissions from any oil and gas exploration or production wells, pipeline compressors or pump stations when determining whether a source is major.

An ''HAP'' is now defined [Section 112(a)(6).] as any ''air pollutant listed pursuant to ''Section 112(b). In Section 112(b), Congress established an initial list of 189 hazardous air pollutants. These listed chemicals are to be viewed as initial candidates for regulation under Section 112.

1. Section 112(b)(2) requires EPA to review periodically the list established in Appendix A and to add or subtract pollutants by rule.
2. A pollutant may be added to the list if ''through inhalation or other

routes of exposure'' EPA deems it worthy of regulation because it presents, or may present, ''adverse human health effects''.

3. ''Adverse human health effect'' has a two-prong definition.
   (a) Under Section 112(b)(2), EPA may list ''substances which are known to be or may reasonably be anticipated to be carcinogenic, mutagenic, teratogenic, neurotoxic, which cause reproductive dysfunctions or which are acutely or chronically toxic'';
   (b) In addition, Congress adopted EPA's definition of ''carcinogenic effect'' from the agency's guidelines for Cancer Risk Assessment at 51 Fed. Reg. 33992, September 24, 1986. [Section 112(a)(11).] This definition may only be modified by notice and comment rulemaking. However, see discussion, *infra*, on residual risk standards.
4. EPA may also list pollutants which present or may present, ''adverse environmental effects''. This latter term is defined as ''any significant and widespread adverse effect which may reasonably be anticipated to wildlife, aquatic life, or other natural resources, including adverse impacts on populations of endangered or threatened species or significant degradation of environmental quality over broad areas''. [Section 112(a)(7).]

The definition of source categories is given as:

A. EPA published a list of categories and subcategories for major and area sources of HAPs. EPA must revise the list no less often than by November 15, 1999, and every eight years thereafter. EPA, to the extent practicable, is required to adhere to pre-existing source categories under Section 111 and the PSD Program in Section 169A(7). [Section 112(c)(1).]
   1. After listing the sources, EPA is required to establish emission standards in accordance with the schedules in Section 112(c) and 112(e). [Section 112(c)(2).]
   2. With regard to categorizing area sources, EPA was to list 90 percent of the area sources that emit the 30 HAPs that present the ''greatest threat to public health in the largest number of urban areas . . .''. [Section 112(c)(3).] Emission standards are to be adopted no later than November 15, 2000.
   3. For sources of alkylated lead compounds, polycyclic organic matter, hexachlorobenzene [sic], mercury, polychlorinated biphenyls, 2,3,7,8-tetrachlorodibenzofurans and 2,3,7,8-tetrachlorodibenzo-*p*-dioxin, EPA must categorize and subcategorize

sources of at least 90 percent of the aggregate emissions of each pollutant by November 15, 1995. [Section 112(c)(6).] Emission standards are to be adopted no later than November 15, 2000.

B. Special Provisions.

1. EPA is not required to, but may, at its option, regulate electric utility steam generating units which emit the following compounds: alkylated lead compounds, polycyclic organic matter, hexachlorobenzene, mercury, polychlorinated biphenyls, 2,3,7,8-tetrachlorodibenzofurans and 2,3,7,8-tetrachlorodibenzo-*p*-dioxin. [Section 112(c)(6).]
2. Research facilities and laboratories may be treated as separate categories "to assure . . . equitable treatment of such facilities." [Section 112(c)(7).]
3. Boat manufacturing is entitled to separate treatment as a source category for styrene unless EPA finds it is inconsistent with the Act's goals and requirements. [Section 112(c)(8).]
4. By implication, the following types of sources may have to be regulated as separate source categories or subcategories.
   (a) Coke ovens; [Section 112(d)(8).]
   (b) POTWs; [Section 112(e)(5); Section 112(n)(3).]
   (c) Radionuclide emissions from NRC licensed facilities; [Section 112(d)(9).]
   (d) Hazardous waste treatment, storage and disposal facilities regulated under RCRA. [Section 112(n)(7).]

C. Delisting of a Source Category.

1. A source category of a carcinogenic HAP may be delisted, if the source category will not emit a HAP pollutant in quantities which may cause a lifetime risk of cancer greater than 1×10^{-6} to an individual who is most exposed. [Section 112(c)(9)(B)(i).]
2. A source category of noncarcinogens or of HAPs that cause adverse environmental effects may be deleted where EPA determines that "emissions from no source in the category or subcategory concerned (or group of sources in the case of area sources) exceed a level which is adequate to protect public health with an ample margin of safety and no adverse environmental effect will result from emissions from any source (or from a group of sources in the case of area sources)". [Section 112(c)(9)(B)(ii).]
3. Delisting petitions must be granted or denied within one year.

The CAA of 1990 established MACT standards as:

A. A maximum achievable control technology (MACT) emission standard is a technology standard defined as "the maximum degree of reduction in emissions of the . . . [HAPs] . . . that the Administrator, taking into consideration the cost of achieving such emission reduction, and any non-air quality health and environmental impacts and energy requirements, determines is achievable for new or existing sources in the category or subcategory to which such emission standard applies, through application of measures, processes, methods, systems or techniques . . .". [Section 112(d)(2).]

1. EPA is to engage in an MACT review for each HAP.
2. In setting MACT, EPA is also to consider any combination of:
   (a) Process changes or substitution of materials; [Section 112 (d)(2)(a).]
   (b) Enclosure of systems or processes; [Section 112(d)(2)(B).]
   (c) Capture and treatment of fugitive, process, stack and storage emissions; [Section 112(d)(2)(C).]
   (d) "Design, equipment, work practice, or operational standards (including requirements for operator training or certification) as required in Section 112(h)". [Section 112(d)92)(D).]

3. The measures described at 2., above, must not, consistent with Section 114(c) of the Act, "compromise any United States patent or United States trademark right, or any confidential business information, or any trade secret or any other intellectual property right."

B. Section 112(d)(3) establishes minimum MACT standards for new and existing sources.

1. For a new source, the MACT standard "shall not be less stringent than emission control that is achieved in practice by the *best controlled similar source*."
2. Existing sources may be regulated less stringently than new sources, but an existing source MACT standard may not be less stringent than:
   (a) The average emission limitation achieved by the best performing 12 percent of existing sources excluding certain LAER sources. LAER is the most stringent technology standard under the Clean Air Act and it applies in nonattainment areas for criteria pollutants.
   (b) LAER standards, prevailing at the time, if 30 or more sources in a category or subcategory area meeting LAER;
   (c) The average emission limitation achieved by the best perform-

ing five sources if the category or subcategory has fewer than 30 sources. [Section 112(d)93)(B).]

3. MACT for an area source category may be "generally available control technologies or management practices." [Section 112 (d)(5).]
4. EPA must review and revise MACT standards, as necessary, "no less often than every 8 years." [Section 112(d)(6).]
5. A NESHAPS standard does not preempt any more stringent standard under Section 111, the PSD Program, the nonattainment program or any state standard. [Section 112(d)(7).]

2.1.4 SUMMARY

Critical technical considerations for the control of air toxics under the 1990 CAA are summarized as:

(1) *Major Source Definition*–All vents, releases and fugitives included in aggregate emissions.

(2) *MACT*–Also requires enclosure, capture, substitution and other process changes.

(3) *Residual Risk*–Controlled source may no longer be acceptable! Ground level impact may be more critical.

(4) *State Risk Criteria*–Additional risk criteria may be applicable, as for example residual risk considerations:
   (a) Criteria: Ground level concentration at property line must meet "AAC" (Acceptable Ambient Concentrations)
   (b) Considerations: Stack height, stack diameter/gas flow velocity, stack location/adjacent structure "downwash," property buffer zones, process location on-site, possible add-on controls

Table 2.2 details possible responses for industries affected by the various provisions of the 1990 CAA Amendments.

2.2 OPERATING PERMIT PROGRAM

One of the most significant parts of the 1990 CAA Amendments is a new general operating permit program established in Title V of this act. An overview of this program includes the following items:

- Applicability: Who Is Subject to the Program?
- State Implemented Title V Air Operating Permit Program Elements
- Permit Fee

- State Program Approval
- Small Business Compliance Assistance Program
- EPA Title V Permitting Rules Published

Some of these items are discussed in more detail below.

The operating permit program will require about 30,000 major sources of regulated pollutants to submit permit applications and may eventually require hundreds of thousands of smaller sources eventually to do likewise. To cover cost of processing these applications at the state agencies, a permit fee is incorporated. These are:

- nominally $25 per ton of each regulated air pollutant
- based on actual emissions
- annual Consumer Price Index (CPI) adjustment
- to be used only to support those elements of the air program listed in Title V of the Clean Air Act

This operating permit program is applicable to:

- major stationary sources of air pollution
- any source required to have a PSD or nonattainment area construction permit
- any source subject to NSPS
- any source (including area sources) subject to the standards for HAPs
- any other source designated by the EPA Administrator under 40 CFR 70
- adequate public participation procedures (notice of proposed actions to citizens and adjacent states, and appropriate judicial review procedures)
- allow minor changes within an affected facility, without a Title V permit revision, if the changes are not "modifications" and do not exceed the emissions allowable under the permit
- general permits and single facilitywide permits allowed
- application shield and permit shield provisions
- adequate permitting authority
  - 5 year permits
  - compliance with SIP limits
  - reopen permits for cause
  - At least $10,000 per day penalty authority
  - criminal penalty authority
  - recognize the EPA administrator's veto authority, etc.
- reasonable and adequate permit processing procedures
- no default permits

The state implemented Title V operating permit program includes the following elements:

TABLE 2.2. **Possible Industry Responses to Air Toxics Provisions of 1990 Clean Air Act Amendments.**

Area	Possible Response
Overall	Determine probable impact of act on facility (e.g., MACT, new permits, accidental release, interaction between MACT and nonattainment portions)
	Develop accurate, up-to-date baseline information on facility operation and configuration
	Review current facility permit (existing sources) and corporate plans for expansion (new/modified sources)
	Determine probable new permit conditions (from both new federal and state/local requirements)
	Determine timing and cost of probable new requirements
	Evaluate strategic options (e.g., MACT or extensions through voluntary 90%/95% reduction, credit for previous BACT/LAER installation)
	Plan for new permits as needed
	Select/implement strategy for corporate response
	Evaluate need to be active participant in EPA rule making
Source prioritization	Determine which of 189 listed chemicals are emitted by facility and how much; develop/refine emission inventory (1991)
	Monitor/comment on possible additional EPA listing of chemicals (1992 and beyond)
	Identify probable source category of facility (ASAP, 1991 – 1992)
	Assess facility health risk (ASAP, 1991 – 1992)
	Determine whether risk-based argument can be made not to list facility's source category
	Monitor/comment on EPA's proposed source category listing (1991 – 1992)
	Estimate probable ranking of facility's source category (ASAP, 1991 – 1992)
MACT compliance	Esimate facility's probable compliance schedule (by 11/15 of following years: first 40—1995, top 25%—1997, top 50%—2000, all 100%—2003)
	Identify and evaluate engineering options for achieving probable MACT and their likely cost (schedule depends on facility category's ranking)
	Evaluate costs and effect of extension options in the act (e.g., voluntary 90%/95% reductions)
	Evaluate implications for corporate planning
	Select/implement strategy for corporate response
Residual risk	Monitor/participate in risk assessment methodology development/refinement (e.g., NAS study, 1996 EPA report to Congress) (now through 1996)
	Assess extent to which residual facility risk after MACT implementation may exceed risk threshold in act (10^{-6})
	Assess extent to which facility risk impact is already regulated (e.g., state/local regulations such as California's Proposition 65 or Air Toxics "Hot Spots" Program (ASAP)
	Prepare for residual risk standards as needed (8 – 9 years away, but necessary preparation should begin much earlier to assure adequate lead time for preparatory work or to identify/evaluate corporate planning options now)

- standard permit application form, and completeness criteria
- monitoring and reporting requirements
- an annual emissions fee requirement
- adequate funding and personnel to administer the program

Each state program must be approved. The CAA requires states to request program delegation from the EPA by November 1993. If the state program is inadequate, EPA is to notify the Governor. State must resubmit request within 180 days of EPA's notice. If the state program is disapproved by EPA, a countdown to mandatory sanctions begins. The EPA administrator may immediately apply any of the sanctions without a grace period. After 18 months, the administrator must apply the sanction. Sanctions include: Highway funding cutoffs & a more stringent emissions offset ratio requirement in nonattainment areas. If an EPA approved state program is not in place by November 1995, EPA must collect the annual fees and administer the program in that state.

2.3 AIR MODELING

Health risk assessments and air dispersion modeling play a key role in various provisions of the Clean Air Act Amendments of 1990. These regulations represent the first major federal requirement to assess public health risk associated with industrial source emissions. It is anticipated that the methodology established in these regulations will be used as a basis for the residual risk standards required by the new air toxics program. Although health assessment and data will be refined in the years leading to promulgation of the residual risk standards, the basic concepts will likely be the same as those considered in the BIF regulations. A discussion of this conceptual methodology is included here in order to illustrate the role of air dispersion modeling in complying with residual risk standards.

The methodology used distinguishes between carcinogenic and noncarcinogenic compounds. For carcinogens, EPA has defined unit risk factors to calculate the estimated incremental lifetime cancer risk to the potential most exposed individual (MEI). The concept is to

(1) Calculate the incremental risk by predicting the maximum annual average ground level concentration for each carcinogenic pollutant
(2) Calculate the estimated risk from that ambient concentration using the unit risk factor
(3) Sum the risk for all carcinogenic pollutants

In this methodology, the MEI is assumed to be located at the point of the maximum annual average ground level concentration. Thus, the MEI risk is the *worst* case risk to any individual, *not* the total population risk. The unit risk factors currently developed by EPA are based on carcinogenic potency factors for humans exposed to known and suspected human carcinogens. These factors are actually representative of the maximum ambient concentrations which result in a "reasonable acceptable" risk (currently defined by EPA as less than one cancer case per 100,000 people). The maximum annual average ground level concentration is determined by conducting an air dispersion modeling analysis.

For noncarcinogens, the methodology is somewhat simpler in that EPA specifies a reference air concentration (RAC) for each compound. This RAC represents the maximum ambient concentration resulting in an acceptable risk as defined by EPA. The RAC is simply compared to the maximum annual average ground level concentration, as determined through dispersion modeling, to determine compliance. To determine compliance with residual risk standards, the following procedure is used:

(1) Predict maximum annual average ground level concentrations.
(2) Calculate risk.
  (a) *Carcinogens*—Multiply maximum concentrations by unit risk factors developed by EPA to calculate unit risk for each carcinogen; sum unit risks and compare to acceptable risk level.
  (b) *Noncarcinogens*—Compare maximum concentrations to acceptable levels defined by EPA.

In summary, methodologies for assessing risk for both carcinogens and noncarcinogens consist in essence of two parts: 1) a factor or quantity defined by EPA representing the concentration of a compound associated with an acceptable risk level, and 2) air dispersion modeling results. It is in the dispersion modeling analysis that the characteristics of the pollution source come into play.

Conducting an air dispersion modeling analysis involves the use of a computer program to perform screening level estimating techniques or actual model simulations. The purpose of both methods is to predict the behavior of the exhaust plume emitted by the pollution source and how the pollutants contained in the plume become dispersed. However, the screening level analysis provides worst case, conservative results as compared to the more realistic model simulation. The model simulation is a much more extensive analysis and requires the use of area specific meteorological data typically for periods of 1 to 5 years.

Currently the computer programs Screen and PTPLU are widely used to conduct screening level analyses. The use of screening level analyses generally results in a more restrictive emission limit. If a model simulation is used, a less conservative limit can usually be established to show compliance with risk-based standards, simply because the analysis is more rigorous.

The Industrial Source Complex Short Term (ISCST) and Industrial Source Complex Long Term (ISCLT) programs are commonly used to conduct simulation modeling. Complex I is widely used in cases where complex terrain exists, to illustrate just how different the results can be when using a program of this type versus a screening analysis.

As part of the RCRA regulations for boilers and industrial furnaces (BIF), EPA has defined a three-tiered approach for assessing compliance with risk-based standards. Tiers I and II are based on a simplified approach to dispersion modeling analysis involving the use of screening tables. The tables were developed by using a computer modeling program to generate maximum annual concentrations for a wide range of source characteristics. To use these tables, the applicant must first define several source characteristics which affect dispersion, including stack height, exhaust gas flow rate and temperature. Tier I tables provide maximum feed rates for each compound to comply with acceptable risk levels. Tier II tables provide maximum emission rates for each compound to comply with acceptable risk levels.

Tier III requires the applicant to develop a site-specific dispersion modeling analysis using a model simulation program such as ISCST. This constitutes a more detailed approach than Tier I and Tier II; however, the results are considered more realistic. There is a built-in conservatism to the Tier I and Tier II approach due to the assumptions used in the methodology to develop the screening tables. Tier III allows the applicant to use assumptions which more closely relate to the actual source characteristics and area meteorology. A summary of the procedure for assessing compliance with BIF risk-based standards is:

Tier I – Feed rate screening tables
Tier II – Emission rate screening tables
Tier III – Site-specific model simulation using ISCST

It is generally anticipated that the residual risk standards eventually promulgated by EPA will be similar to the BIF methodologies and strategies. Because of the wide range of industries affected by the Clean Air Act Amendments of 1990, the standards will have to address a variety of source types. Risk standards will be required for a long list of pollutants, thus establishing the need for more extensive health effects data. The role of dispersion modeling will be greatly increased as more and more facilities are required to assess air quality impact.

In summary, dispersion modeling analyses show either or both:

(1) Screening analysis – yields conservative results

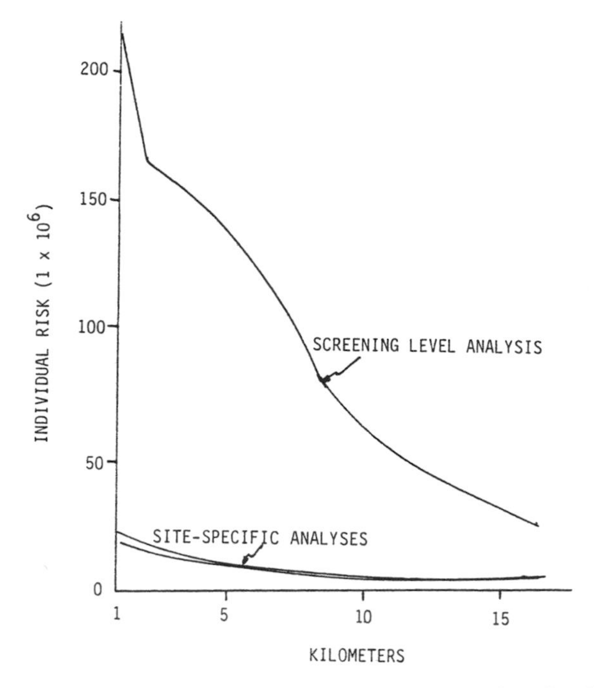

Figure 2.1 Comparison of risk for two different screening techniques: low-elevation plume.

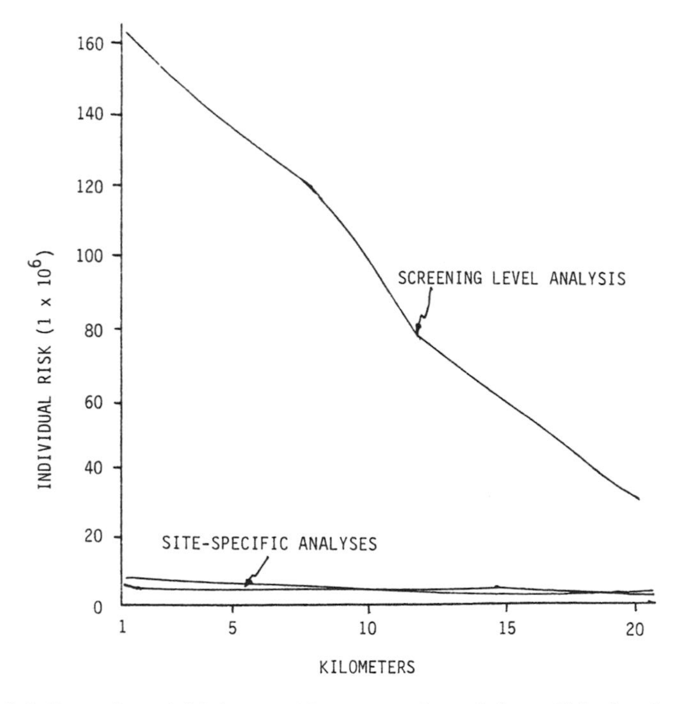

Figure 2.2 Comparison of risk for two different screening techniques: high-elevation plume.

    (a) EPA screen
    (b) PTPLU
  (2) Model simulation – yields more realistic results
    (a) ISCLT, ISCST
    (b) Complex I

An example of screening level analyses generated by PTPLU compared to site-specific analyses generated by ISLST is shown in Figures 2.1 and 2.2. The differences are dramatic.

CHAPTER 3

The Use of Concept Engineering Reports

FRANK L. CROSS, JR., P.E.[1]

3.1 INTRODUCTION

CONCEPT reports can be useful in sizing and selecting control systems. To show this, an example of a study on a biohazardous waste incineration system will be used. The purpose of this study was to evaluate the feasibility and develop a conceptual design for a new waste incineration facility. This study includes sections on waste characterization, available incineration and air pollution control technologies, environmental regulations, system housing, system description and economics.

3.2 WASTE CHARACTERIZATION

The quantity and type of waste must be known in order to select the proper incinerator equipment size. A waste survey was conducted at the facility in order to provide the necessary waste characterization for an incineration system. The survey was conducted by C/TA[1] and personnel. A detailed description of the survey methods and findings can be found in the previously submitted waste survey report.

Waste amounts and types were quantified as a result of the waste survey and the results are given in Table 3.1. The facility currently generates approximately 2500 pounds per day. This is expected to climb to an estimated 3900 pounds per day, after the proposed laboratory buildings are completed. The majority of the waste is animal bedding and carcasses. Only a small percentage, approximately 10%, is red and orange bag waste.

[1]President, Cross/Tessitore & Associates, P.A., Orlando, Florida.

TABLE 3.1. Waste Survey Results.

Waste Types	Current Generation Rate (lb/day)	Future Generation Rate (lb/day)	Portion of Wastes Stream (%)
Orange bags	130	310	8
Red bags	40	60	2
Bedding	1550	2330	59
Carcasses	700	1050	27
Miscellaneous	110	170	4
Total	2530	3920	100

An estimate of the waste's energy content is made once the types and quantities of waste are known. This estimate is arrived at by using the results of the waste survey along with published information and past experience. The results are given in Table 3.2. The energy content of the waste along with the quantity of waste is then used to help perform a mass and energy balance.

The proposed incinerator should be able to handle 2500 pounds of waste per day now, and 4000 pounds of waste per day in the future, with an average waste energy content of 4400 Btu's per pound. An incinerator sized to handle 400 to 500 pounds per hour and operated 8 to 10 hours per day would easily accommodate the projected waste load. This size incinerator would be more amenable to a high efficiency air pollution control system than a smaller incinerator. This scenario would also reduce labor costs compared to the current 18-hour per day operating schedule.

3.3 DISCUSSION OF AVAILABLE TECHNOLOGIES

This section discusses and compares the technologies currently available for incineration and air pollution control.

3.3.1 INCINERATION

The two most common technologies for medical waste incineration are rotary kiln and controlled air systems. One variation of the controlled air system is the pathological incinerator. These three systems are discussed below.

3.3.1.1 Rotary Kiln

The rotary kiln incineration system consists of a long, cylindrical, rotating chamber followed by a fixed combustion chamber, as shown in Figure

3.1. The rotating chamber allows the solid waste to be tumbled during combustion, while the fixed secondary chamber provides for destruction of any volatilized organic compounds. The first stage tumbling action facilitates proper mixing of waste and air that is essential for good combustion and subsequent "burnout" of organic material contained in the waste.

Traditionally, rotary kiln units have been the most reliable for systems requiring continuous operation, automatic ash removal, and heat recovery. However, particulate emissions are not controlled in any way, thus, air pollution control equipment is usually required to meet even the most lenient regulations. This tends to increase the cost of the system. In addition, capital costs for rotary kilns are typically 1.5 to 2 times greater than controlled air units of comparable size. Maintenance costs are typically greater for rotary kiln systems as well. This type of incinerator is not usually used except for regional systems because of its size.

3.3.1.2 Controlled Air

The controlled air incinerator consists of dual combustion chambers, a fixed primary and a fixed secondary. The controlled air unit gets its name from the fact that airflow in the primary chamber is regulated in order to minimize the amount of particulate entrainment in the exhaust gases. In the primary combustion chamber, waste is fed onto a hearth and burned under substoichiometric combustion conditions. This process is sufficient to volatilize organic compounds and, because the airflow is relatively small, particulate matter tends to remain on the chamber floor. The gaseous products of this combustion are drawn through to the secondary chamber where additional burning takes place utilizing excess air. This ensures complete combustion of organic compounds. In order to achieve consistent and complete burning of all waste types, a burn down cycle should be included in the daily operating schedule of cyclic units. Some problems have traditionally been associated with the operation of controlled air incinerators

TABLE 3.2. Energy Content of Waste.

Waste Type	Generation Rate (lb/day)	Estimated Energy Content (Btu/lb)	Total Energy (Btu/day)
Red/orange bags and miscellaneous	280	9400	2.63×10^6
Bedding	1550	5000	7.75×10^6
Carcasses	700	1000	0.7×10^6
Total	2530		11.08×10^6

Average energy content of mixed waste—11.08×10^6 Btu/day – 2530 lb/day = 4379 Btu/lb.

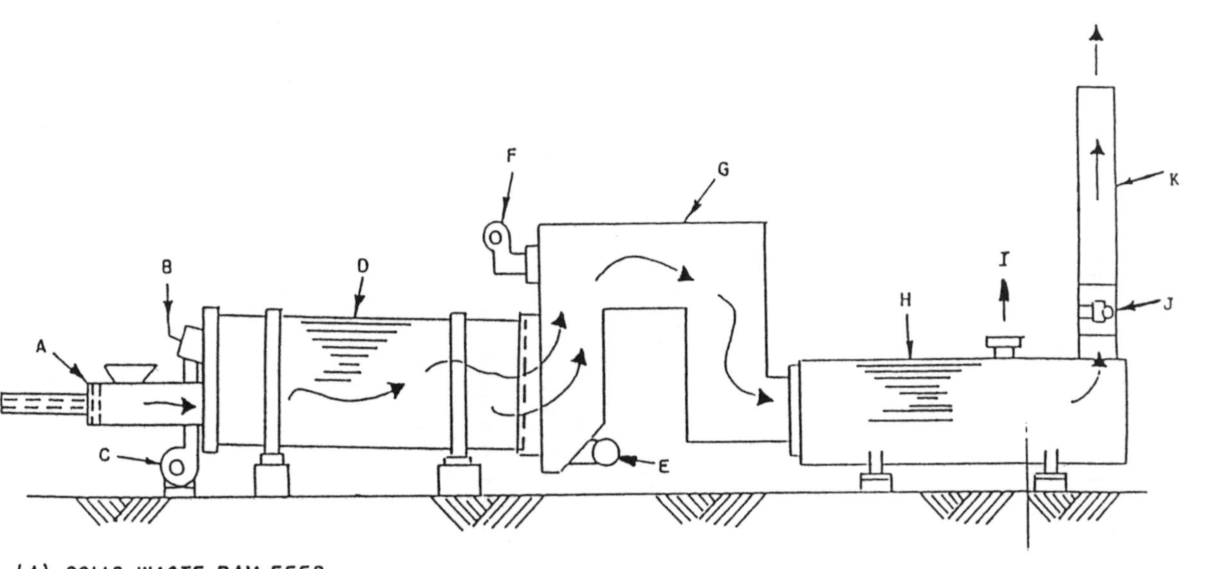

(A) SOLID WASTE RAM FEED
(B) LIQUID WASTE AND AUXILIARY FUEL BURNER
(C) COMBUSTION AND DILUENT AIR BLOWER
(D) ROTARY KILN INCINERATOR
(E) ASH REMOVAL
(F) AFTER BURNER LIQUID WASTE AND AUXILIARY FUEL
(G) AFTER BURNER SECTION
(H) WASTE HEAT RECOVERY BOILER
(I) STEAM OUTLET
(J) EXHAUST BLOWER
(K) EXHAUST STACK

Figure 3.1 Typical co-current rotary kiln incinerator.

on a 24-hour continuous basis. Design modifications have been implemented by certain manufacturers to increase the performance level of continuous duty systems. A typical controlled air incinerator is shown in Figures 3.2 and 3.3. The controlled air incineration technology is generally much more cost effective for a system of the size considered in this study.

3.3.1.3 Pathological

A pathological incineration is a variation of the controlled air incinerator. Several modifications are made to account for the high moisture content and low heating value of pathological waste. These units are designed to handle pathological waste (including animal bedding and carcasses) and are generally unsuitable for any other waste types.

A pathological incinerator must incorporate design features to vaporize the large amount of moisture present in the waste. These design features provide a large amount of heat and a restricted supply of air to aid in the vaporization process. The heat is usually provided by larger burners in the primary chamber or by placing the secondary chamber directly underneath the primary chamber (see Figure 3.4). Air is introduced to the primary chamber through a series of air ports above the hearth.

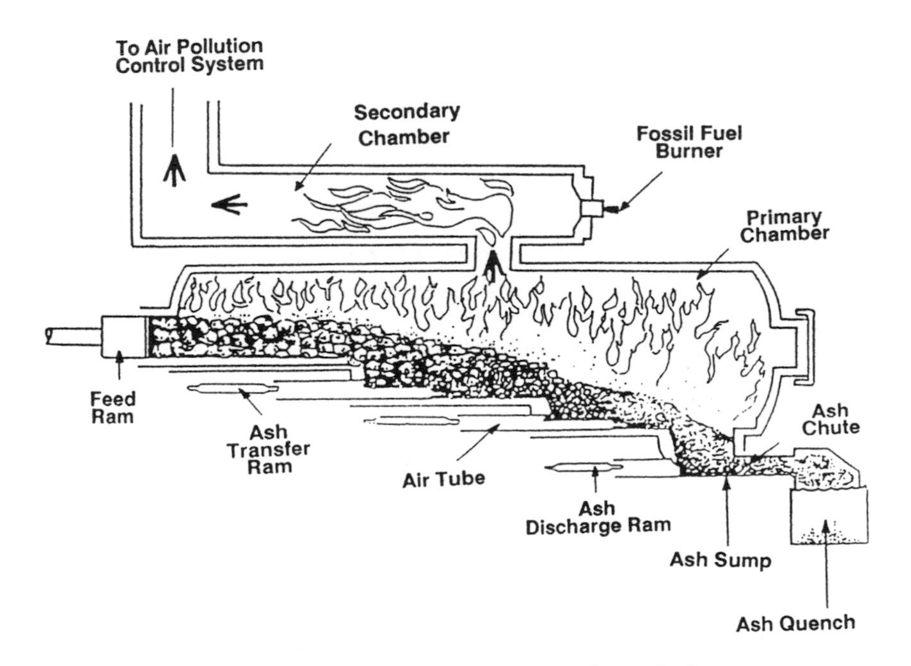

Figure 3.2 Controlled air incinerator (continuous duty).

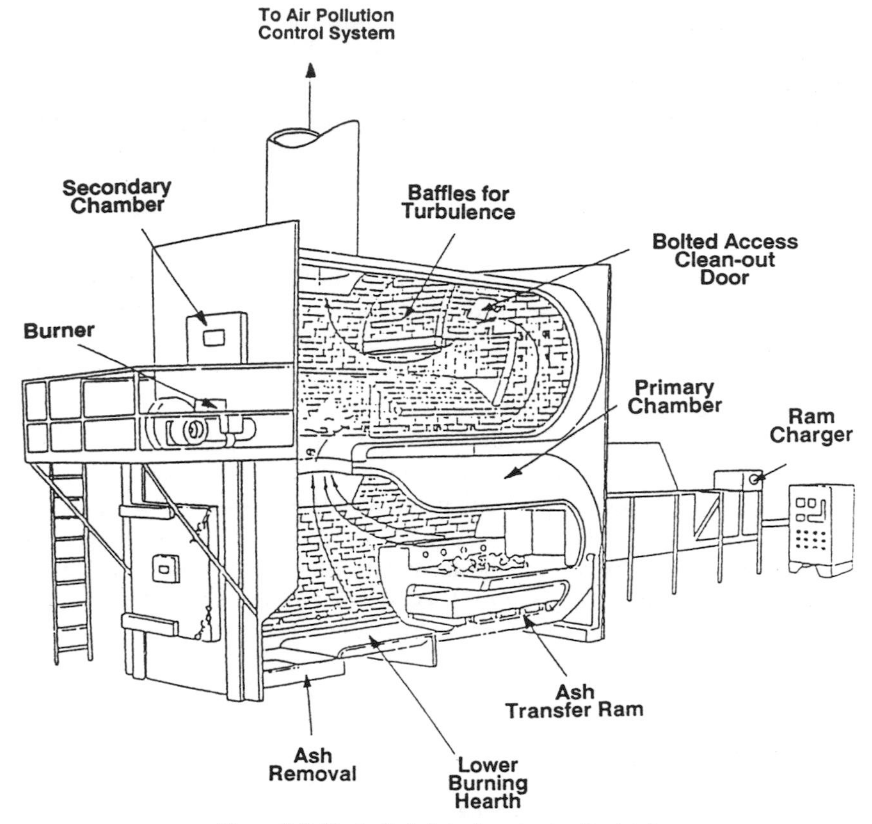

Figure 3.3 Controlled air incinerator (cyclic duty).

A pathological incinerator has a specially designed hearth. Liquids, such as oils and fats, are often released during the combustion process. The hearth should be designed to contain these liquids and prevent them from flowing out of the combustion chamber.

The feed system of a pathological incinerator may also differ from a general use controlled air unit. The general use unit would have a ram feeder which pushes the waste into the primary chamber from the side. A pathological unit is likely to have a top loader, where a chute extends from the top of the combustion chamber. Waste is fed into the chute and falls into the primary chamber. This helps prevent disfiguring of carcasses and the resulting dissemination of body fluids and tissue. The distinction between a general use controlled air unit and a pathological unit is not always clear. Manufacturers will often incorporate the features mentioned above to varying degrees. The resulting incinerators are intended to accommodate mixed waste streams.

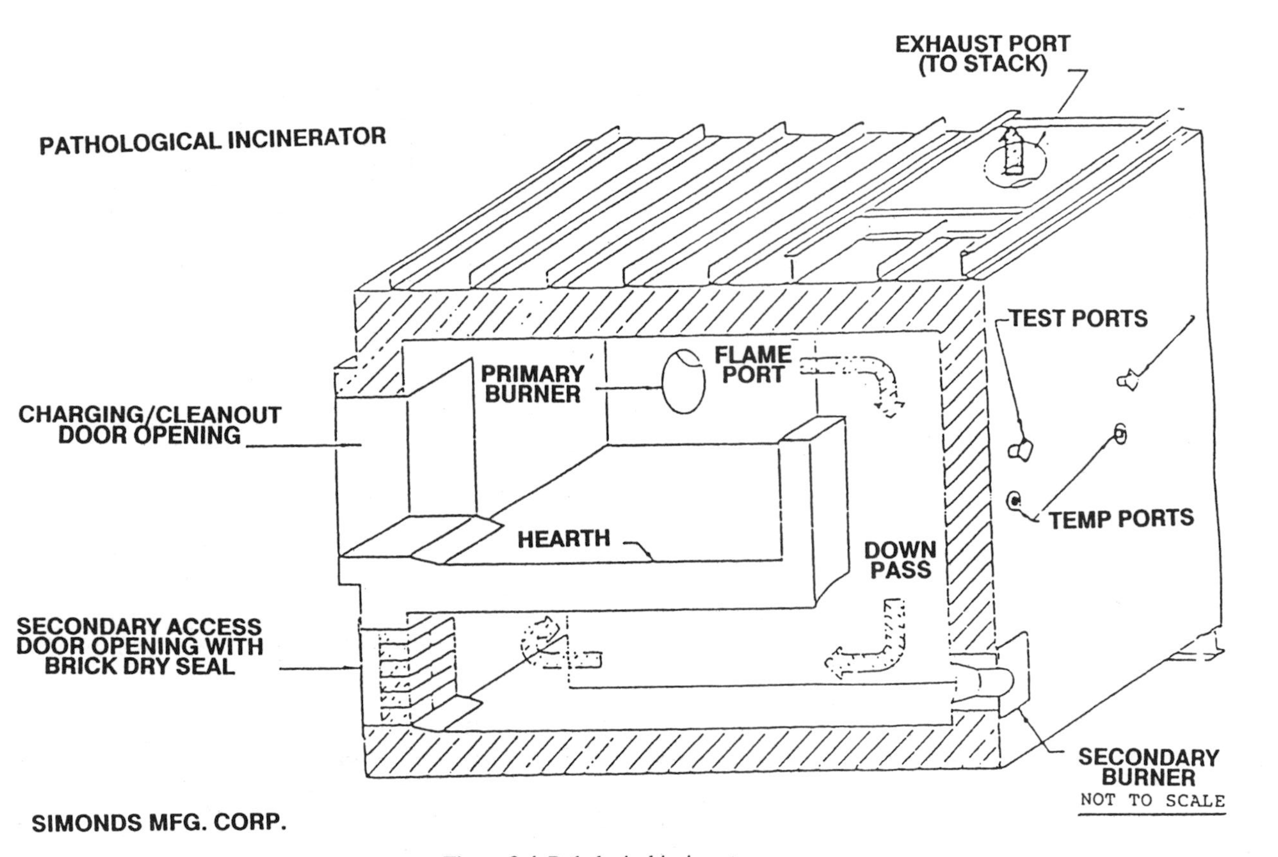

Figure 3.4 Pathological incinerator.

3.3.2 AIR POLLUTION CONTROL

Environmental regulations currently imposed on biohazardous waste incineration necessitate the addition of air pollution control equipment.

The following paragraphs provide a discussion of four highly efficient air pollution control technologies which would satisfy these requirements.

3.3.2.1 Venturi-Packed Tower

The Venturi-packed tower scrubber control system consists of two separate stages, as shown in Figure 3.5. The first stage, a Venturi scrubber, removes particulates from the gas stream, while the second stage, the packed tower scrubber, removes gaseous contaminants and remaining particulate from the exhaust gas stream.

Fresh water is injected into the high velocity gas stream either at the inlet to the converging section of the scrubber, or at the Venturi throat. The kinetic energy of the gas stream atomizes the liquid into droplets, maximizing the surface area of the liquid. The resulting gas/liquid contact permits removal of particulate contaminants by adsorption.

The energy needed to accelerate the gas stream through the Venturi throat is supplied by an induced draft fan. To maintain the required high velocity gas stream for efficient particulate removal requires a large pressure drop across the Venturi throat. This entails a large induced draft fan with significant power consumption. When particulate emission limits become more stringent, a greater pressure drop across the Venturi throat is required to meet the limits. The greater pressure drop means increased electrical power consumption by the induced draft fan. The ability of a Venturi system to meet possible future particulate emission limits and retain reasonable operating cost is uncertain (i.e., 0.03 gr/scf is achievable, but 0.015 gr/scf is unlikely).

Following scrubbing in the Venturi, the solids are separated from the gas stream in a bottom fed cyclone, which is the lower section of the packed tower. Remaining gaseous pollutants are removed in the upper packed section of the tower. This is generally accomplished through a counter flow packed tower arrangement. The packed tower scrubber is a vessel filled with randomly oriented packing materials (called dump packing), such as saddles and rings. The scrubbing liquid is fed to the top of the vessel, and recycled with chemical treatment from the bottom. As the liquid flows down through the tower, it wets the packing material, providing a large surface area for mass transfer with the gas stream. The cleaned gas normally passes through a mist eliminator and is then vented through a stack to the atmosphere. The liquid leaving the absorber is either stripped of the contaminant gas and recycled, or passed on for further waste treatment or process use.

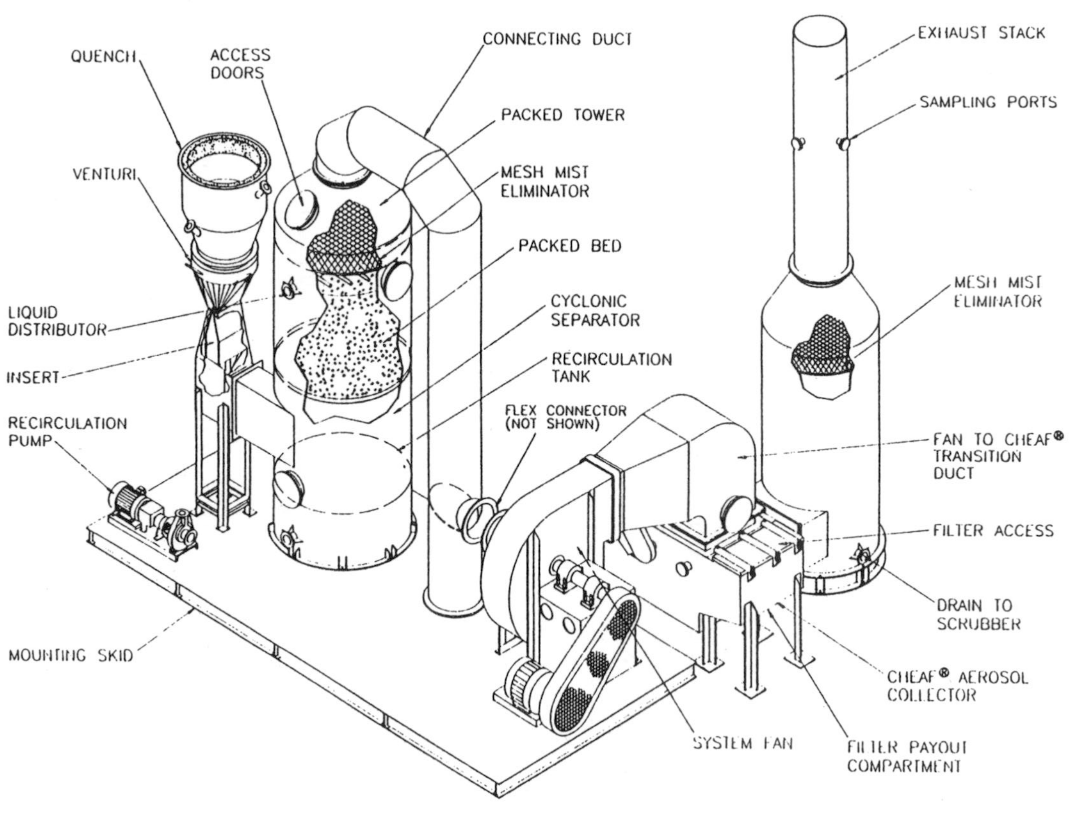

Figure 3.5 Venturi scrubber followed by packed tower.

The packing material (inert chemically) is designed to increase the surface area for gas liquid contact.

The Venturi-packed tower scrubber system is generally lower in capital cost than the other technologies and is a relatively compact system, requiring a minimum of floor space.

Wet scrubbing systems may emit a visible plume from the exhaust stack. This plume is often misinterpreted by the public and casts a negative connotation on the incineration system. This can be avoided by reheating the exhaust gas. The reheating is easily done but represents a substantial additional cost in the form of fuel. Some wet scrubber manufacturers (i.e. Calvert and ACI) take steps to remove moisture from the exhaust stream. This helps reduce the fuel cost associated with reheating the exhaust gases.

Wet scrubber systems will also have a liquid discharge (called a bleed) from the packed tower that will require disposal. This is most often piped to the sewer with no pretreatment. When sewer connections are not available or the local wastewater treatment plant is hesitant to accept the discharge, then the discharge becomes a concern that may affect the equipment selection process.

Some manufacturers of wet scrubber systems offer a zero-liquid discharge option. This option circumvents the aforementioned problems but also significantly increases the capital cost of such a system. Much of the price advantage of a wet scrubber over a dry scrubber is lost when this option is purchased.

3.3.2.2 Dry Scrubber

A typical dry scrubber consists of two main units, a chemical reactor, followed by a baghouse (see Figure 3.6).

The two variations of this system that are commonly used vary in the way the reagent is injected. Spray dryer systems inject the reagent as a slurry, while dry sorbent injection (DSI) systems use a dry powder. Spray dryer systems have traditionally been used in conjunction with large-scale industrial processes. Dry sorbent injection is now emerging as a control technology particularly suited for hospital-size incinerators.

In a DSI system, the flue gas enters a quench chamber. In the quench chamber, the gas is cooled and humidified as a conditioning step for acid gas removal.

The conditioned flue gas then enters a reactor vessel where the reagent, a dry powder, is pneumatically injected. The reagent injection takes place in a high velocity mixing zone, so that mixing of the reagent and gas is maximized. Neutralization of the acid gases occurs as the reagent reacts

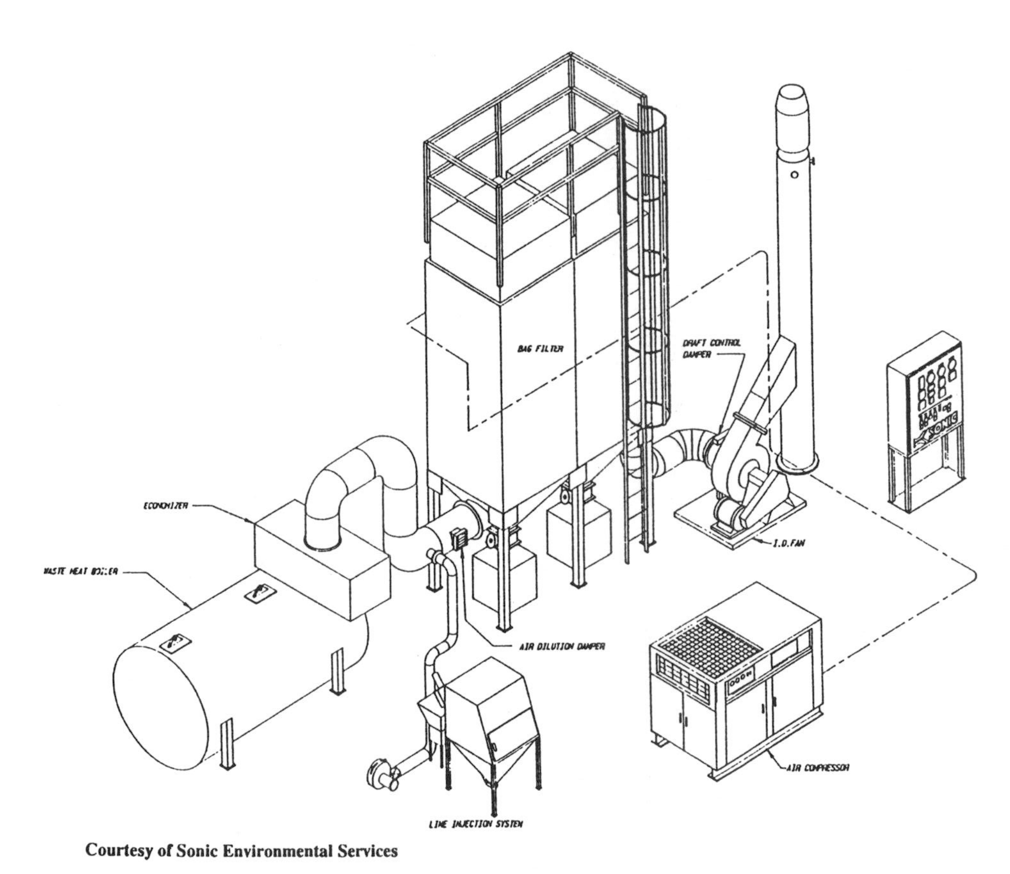

Courtesy of Sonic Environmental Services

Figure 3.6 Dry scrubber (Courtesy of Sonic Environmental Services).

with the acid to form reaction salts and water. A typical reaction is represented as:

$$Ca(OH)_2 + 2HCl \rightarrow CaCl_2 + H_2O$$

The gas is then passed to a baghouse or electrostatic precipitator. Baghouses have generally shown higher removal efficiencies than precipitators. In the baghouse, fabric filters consisting of cylindrical filter bags are vertically suspended in a baghouse compartment. The filter bags are porous which allows the gas to pass through them while capturing a high percentage of the particles suspended in the gas stream. The fly ash, reaction salts, and unreacted reagent are entrained in the gas stream, and are captured in the baghouse as the solids collect on the filter bags. Further acid gas removal takes place on the filter bags as the flue gas passes through the filter cake containing the unreacted reagent. As the particulate cake builds on the bags, the pressure drop across the bags also increases and the bags must be cleaned. Cleaning the bags dislodges the particulate cake that falls into a hopper for removal and disposal.

Dry sorbent injection followed by a baghouse has a number of advantages and disadvantages. There is no liquid discharge from the system, but the residue collected in the baghouse must be disposed of. The water used in the quench chamber is completely evaporated, so there is no worry about a visible steam plume coming from the exhaust stack.

The baghouse is a proven device and is capable of meeting future emission limits with little or no increase in operating cost.

The DSI portion of the system is an emerging technology. Achieving the necessary level of acid gas removal generally requires approximately twice the amount of reagent, as in a packed tower system. Reagent handling problems, such as moisture in the injection system, are a concern. The moisture may cause the reagent to cake up and clog the injection nozzles.

A DSI baghouse system uses a smaller induced draft fan than a wet scrubber system, so utility costs are significantly reduced. Maintenance costs for a baghouse tend to be greater than for a wet scrubber because the fabric filter bags must be changed periodically.

3.3.2.3 Baghouse/Packed Tower

The baghouse/packed tower system includes one component from each of the previously discussed technologies. The purpose of this system is to provide a high level of particulate removal with the baghouse and a high level of acid gas removal with the packed tower. Since both the baghouse and the packed tower are proven and operating components, the combination is considered readily achievable, but no single manufacturer produces

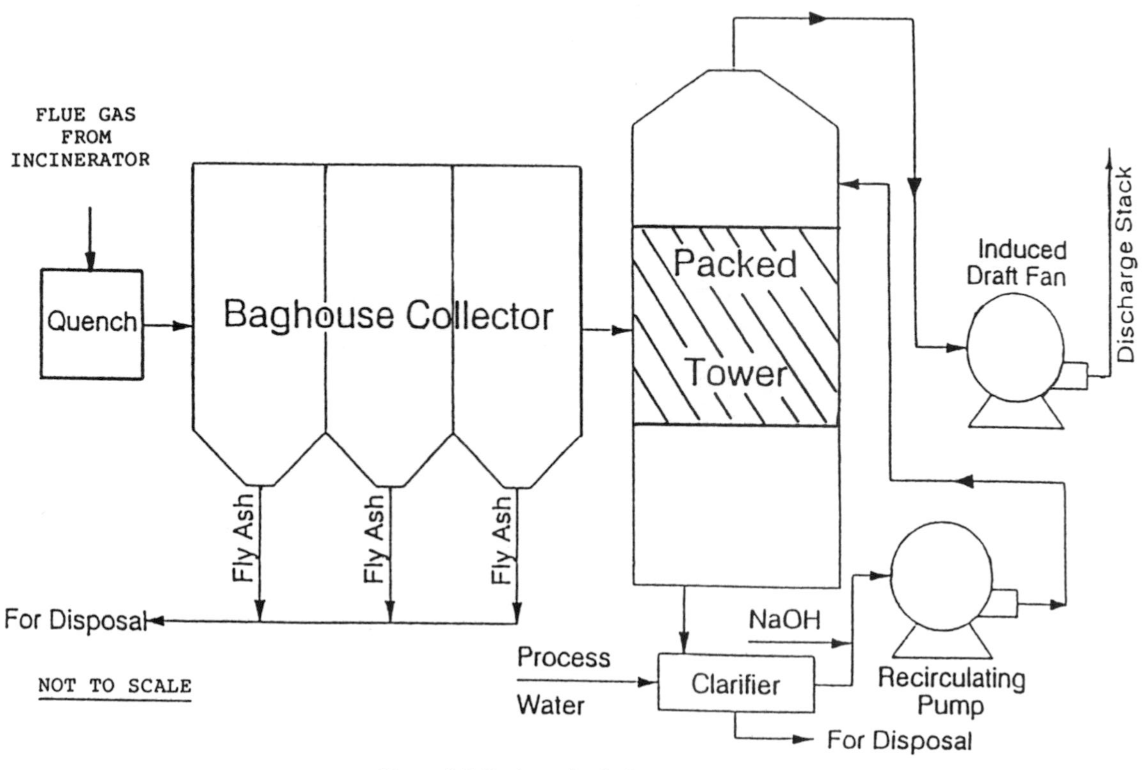

Figure 3.7 Baghouse/packed tower system.

such a system. The baghouse and packed tower would have to be purchased from different manufacturers and then combined. Figure 3.7 provides a diagram of the baghouse/packed tower system.

3.3.2.4 Calvert Collision Scrubber

The Calvert collision scrubber, pictured in Figure 3.8, is a unique design that maximizes the capability of a conventional Venturi scrubber throat and then takes another major step in collecting particles and absorbing gas. The scrubber has three scrubber throats that are arranged like a "T." Gas entering the collision scrubber is split into two gas streams prior to entering a pair of scrubber throats. Water is injected at the entrance of each throat and is atomized by the gas stream. The droplets collect particles as they are accelerated towards the center of the throat.

At the center, the two gas streams collide head-on. The droplets, because of their inertia, travel into the opposing gas stream where they attain a high differential velocity. This collision process shreds the water drops into finer ones that more effectively collect submicron particles and produce a larger liquid surface area for gas absorption.

This type of technology does compare favorably on a cost basis with the previously discussed systems. So far there are only one or two of these systems on a medical waste incinerator. It can be oriented to be a horizontal or a vertical system, depending upon the space available.

Like the conventional Venturi-packed tower, it has a liquid discharge that must be disposed of. Reheating of the exhaust gases may also be necessary to avoid emitting a visible steam plume from the stack. The amount of reheating is generally less for the Calvert than the conventional Venturi-packed tower. Other similar systems (i.e. ACI, etc.) are now being marketed at the present time.

The advantages and disadvantages of the air pollution control systems discussed are listed in Table 3.3.

3.4 ENVIRONMENTAL REGULATIONS

Regulations governing the disposal of medical waste are in a state of flux. The Air Quality Division (AQD) of the Michigan Department of Natural Resources (MDNR) has issued guidelines for permitting medical waste incinerators. The guidelines give the potential permittee an idea of what to expect, but they are not binding. The MDNR handles each permit review on a case by case basis. They have considerable latitude in deciding the conditions for a permit. Meeting with the MDNR in the initial stages of the

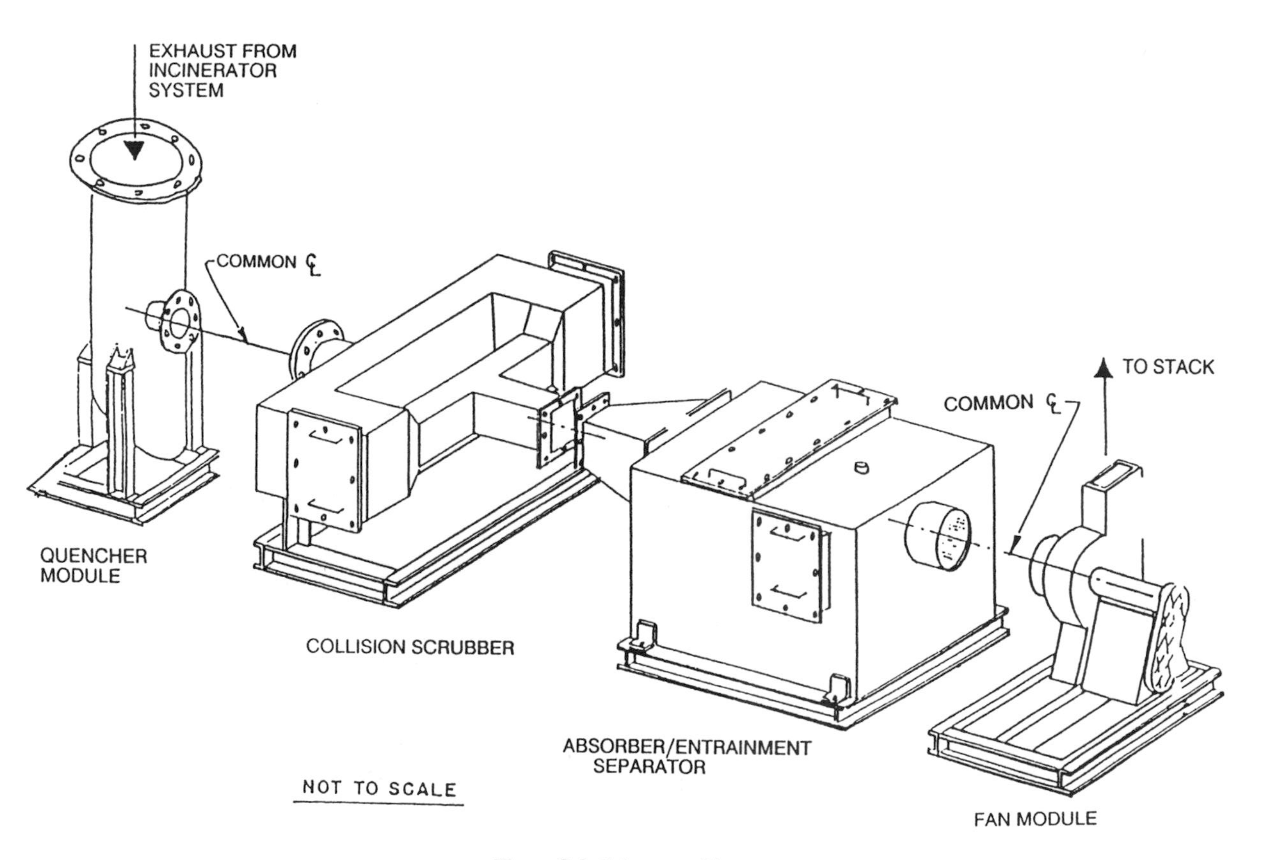

Figure 3.8 Calvert scrubber.

TABLE 3.3. **Air Pollution Control System Comparison.**

	Advantages	Disadvantages
Venturi-packed tower	–Compact system –Low capital cost –Proven technology	–Possible difficulty and increased operating expenses to meet suggested future emission limits –High power (electrical) consumption –Visible steam plume
Dry scrubber	–Lower power (electrical) consumption –No visible steam plume –No liquid discharge –Capable of meeting suggested future emission limits	–Increased reagent consumption –High captial cost –Limited field experience for hospital incinerator applications –Possible reagent handling problems. –Requires more floor space and taller ceiling
Baghouse/packed tower	–Proven technology for individual components	–Not readily available
Calvert collision scrubber	–Minimal steam plume –Reduced power (electrical) consumption compared to Venturi-packed tower	–High capital cost to meet future emission limits

project would be very beneficial. A clearer picture of the requirements will emerge once the MDNR has a good idea of what is involved. Being granted an air permit is one of the key aspects of this project. The permitting process can slow or stop a project of this nature. An upfront meeting with the state will help avoid future difficulties.

Equipment design and selection should not be dictated solely by current regulations. Future regulations should also be considered. This is difficult to do since there is no clear consensus on future regulations. It is important because new regulations are not likely to contain any grandfather clauses for existing units.

The MDNR's current guidelines are summarized in Table 3.4. The guidelines vary depending on the type and amount of waste.

The permitting process itself is quite involved. The needed information is given in Table 3.5. The initial submittal is often followed by one or two additional information requests from the state. It generally takes 6 to 12 months to secure a permit.

Permit conditions generally include a stack test. The stack test is used to demonstrate that the permit emission limits are being met. The test is usually done shortly after the equipment is commissioned and might be required annually or biannially, thereafter.

Disposal of radioactive material by incineration must be approved by the Nuclear Regulatory Commission (NRC). The only exceptions to this are:

(1) 0.05 microcuries or less of hydrogen-3 or carbon-14, per gram of medium used for liquid scintillation counting

(2) 0.05 microcuries or less of hydrogen-3 or carbon-14, per gram of animal tissue averaged over the weight of the entire animal

The Code of Federal Regulations (10 CFR 20.305 and 20.306) discusses the treatment by incineration of this waste. The method for obtaining approval of proposed disposal procedures for other licensed material is addressed in 10 CFR 20.302. Appendix B, of Part 20, lists the concentrations in air above natural background for other radionuclides.

TABLE 3.4. **MDNR Incineration Guidelines.**

	Pathological Units	Medical Waste Units (<450 lb/hr)	Medical Waste Units (>450 lb/hr)
Particulate matter (gr/dscf @ 7% O^2	0.10	0.06	0.015
Hydrogen chloride (ppm @ 7% O^2)	No std.	Based on disperson modeling	50
Opacity (%)	No std.	20	10
Carbon monoxide (ppm @ 7% O^2	No std.	No. std	75
Dioxins, furans and heavy metals	Limits determined on a case by case basis. Best available control technology may be required.		
Secondary condition chamber temperature (°F) (for cytotoxic materials)	No std. —	1800 (2200)	1800 (2200)
Secondary combustion chamber retention time (seconds) (for cytotoxic materials)	No std. —	1 2	1 2
Continuous monitoring	No requirements	Combustion temperature	Combustion temperature Carbon monoxide Opacity

TABLE 3.5. **MDNR Permit Requirements.**

(1) Plot plan of site (including building heights and air intake locations).
(2) Detailed description of the incinerator
(3) Detailed description of the waste
(4) Operating schedule
(5) Combustion chamber temperature and retention time profiles
(6) Description of the temperature monitoring system
(7) Start-up and shutdown procedures
(8) Exhaust gas volume and composition
(9) Description of the combustion controls
(10) Discussion of the feasibility of air pollution control
(11) Complete description of the air pollution control equipment
(12) Description of any bypass of the air pollution control system
(13) Location of stack sampling ports
(14) Description of the ash handling system
(15) Complete description of the maintenance program
(16) Maximum uncontrolled and controlled emission rates for:
  Particulate
  Sulfur dioxide
  Nitrogen oxides
  Carbon monoxide
  Polychlorinated biphenyls (PCBs)
  Mercury
  Arsenic
  Cadmium
  Chromium
  Dioxins
  Furans
  Hydrogen chloride
  Benzo-a-pyrene
(17) Computer dispersion modeling for the above pollutants showing acceptable ambient concentrations

3.5 SYSTEM HOUSING

Preferred layout is for the incinerator to be on the ground floor with the air pollution control on the roof. This could be accomplished in a number of ways using various combinations of equipment. The exact building configuration and design should be decided after the equipment type has been selected. A conceptual sketch is shown in Figure 3.9.

Estimated space requirements for a 400-pound per hour incinerator and pollution control system are given in Table 3.6. These are intended to give some idea of the space required, but they are only preliminary estimates at this point. The dimensions would allow room for equipment and access to the equipment. Any room needed for waste storage or staging has not been included.

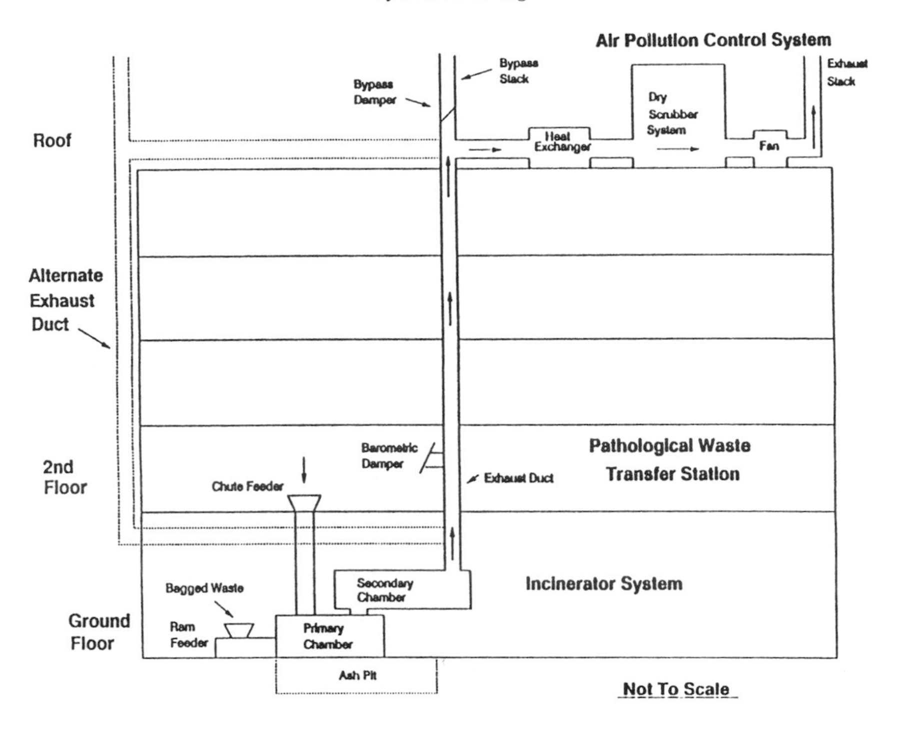

Figures 3.9 Incinerator/APC concept.

Some incinerators have waste feeders designed to be accessed from the floor above. Space on the floor above where the incinerator is housed would have to be available if this model incinerator is chosen. The incinerator operator would also have to be able to move easily from one level to the other.

Utility requirements will vary depending on the type of system selected. The possible utility types and amounts for a 400-pound per hour system are estimated in Table 3.7. Some of these utilities may not be needed, depending on which air pollution control system is selected.

Locating the air pollution control system remotely from the incinerator will require a long exhaust duct to connect the two units. This duct without

TABLE 3.6. Equipment Space Requirements.

	L (ft)	W (ft)	H (ft)
Incinerator	50	20	20
Air pollution control system	25	20	20–35
Ash pit	30–35	12	4–5

TABLE 3.7. **Utility Requirements.**

Incinerator		
	Electrical	460 volt 3 phase, 70 amps
	Natural gas	4000 cfh
	Ventilation	1500 scfm
	Water	4 gpm
Air pollution control system		
	Electrical	460 volt 3 phase, 60 amps
	Water	12 gpm
	Sewer	8 gpm (wet scrubber only)
	Compressed air	100 psi (dry scrubber only)

external insulation will be approximately two to three feet in diameter. Additional insulation will be needed if the duct traverses the interior of the building. The duct could also be routed along the outside of the building, if so desired.

The weight of the air pollution control system may be of concern if the system is mounted on the roof. Estimated operating weights for different types of systems are given in Table 3.8. The building structure may have to be strengthened to accommodate the load. A weatherproof enclosure for the air pollution control system is recommended. This is desirable for aesthetic reasons and for increased equipment life.

3.6 SYSTEM DESCRIPTION

The recommended incineration system utilizes a controlled air incinerator. The use of this incinerator will allow for two modes of operation, one for the burning of pathological wastes, and one for the burning of infectious wastes. The destruction of the pathological waste will employ the full use of the burners and throttling back the air supplied to the primary combustion chamber. In handling the infectious wastes, the burners will be throttled back and more air will be supplied. This incinerator will provide the flexibility to handle the pathological waste for the majority of the shift, reserving a portion at the end of the shift to deal with infectious wastes. The incinerator is sized to handle 400 to 500 pounds per hour, which will handle the current, as well as projected loading. The secondary combustion cham-

TABLE 3.8. **Air Pollution Control System Operating Weights.**

Dry scrubber	90,000 lbs.
Wet scrubber	30,000 lbs.

ber is sized to provide the desired one and one-half seconds residence time (at 1800°F) for good destruction of the volatile gases (see Figure 3.9).

The use of a waste heat to energy conversion system was evaluated. It was concluded that the amount of energy (in the form of steam) that could be recovered was too small to be cost effective for this project.

The recommended incineration system consists of the following components:

- Hydraulic Ram Feed Mechanism
- Top Loading Chute Feed
- Primary Combustion Chamber
- Secondary Combustion Chamber
- Mechanical Ash Handling and Disposal System
- Air Pollution Control System

3.6.1 HYDRAULIC RAM FEED MECHANISM

The ram feed mechanism is a hydraulically operated waste charging device that injects waste into the primary combustion chamber (See Figure 3.10). The ram feed will be used to convey the bagged infectious wastes to the primary chamber. The ram feeder will be accessed from the ground floor. Operation of the ram feed mechanism is controlled by the incineration control system. The operator will place a charge in the hopper. The hatch cover will be closed and the incinerator placed in the "load" mode. The charging door will open and the ram will move forward through the open charging door, pushing the charge into the primary chamber. The ram will move back clear of the charging door and the charging door will close. The ram will reset to the normal position and the hatch cover will automatically open to accept the next charge. Control interlocks will prevent the operation of the ram feeder if the hatch cover is open, the charging door is closed, or the incineration controls are not in a "load" mode. This will provide for both operator and equipment safety, as well as an air lock for the waste charge because the incinerator charging door will not open until the feed hopper cover is closed.

3.6.2 TOP LOADING CHUTE

The top loading chute will provide access to the primary chamber for the loading of pathological waste material. The chute is oriented vertically and accessed from the second floor. This allows for a gravity feed operation that is controlled by three hydraulically operated doors. This arrangement will minimize the possibility of any fumes, smoke, or particulate matter from

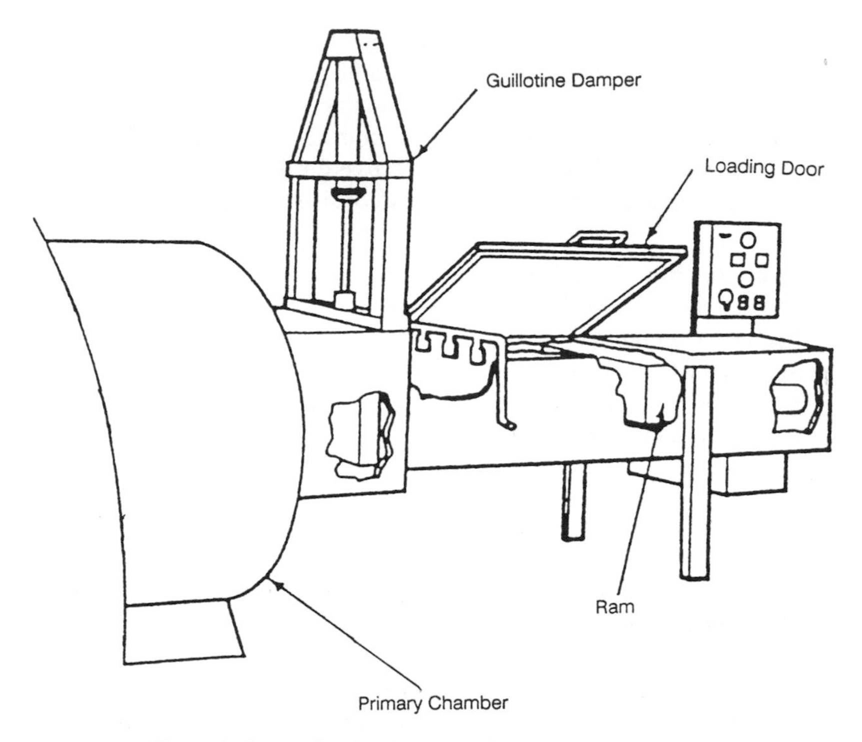

Figure 3.10 Ram feed loading device for controlled-air incinerator.

escaping the incinerator. The loading hatch is gasketed, and a blower will purge the loading chute after each charge.

The operation of the chute will be controlled automatically. The loading hatch will be opened after the chute has been purged, and the charging door and waste support door have been closed and latched. After the charge has been placed on the load support door, the loading hatch will be hydraulically closed and sealed. At the occurrence of a preset time interval, interlocked with temperature controls, both the load support door and the charging door will open simultaneously. Both doors will close and latch after a timed interval for charging. The temperature control interlocks on the loading system prevent the primary chamber from being overloaded with pathological waste. By preventing overloaded conditions in the primary chamber, the temperature is consistently maintained within the set points. The interlocks also safeguard against the primary chamber temperature getting too high.

3.6.3 PRIMARY COMBUSTION CHAMBER

The primary combustion chamber will be a refractory lined chamber with a firebrick hearth area. The chamber is insulated with mineral wool block insulation that is installed between the refractory and an outer steel shell. All of the door openings in the primary chamber are fully gasketed to control fluid leakage, and prevent any gaseous emissions. An air distribution system will provide the required amount of over fire combustion air. The primary chamber will be equipped with two burners that will provide a capability to preheat the chamber, and automatically maintain the temperature between predetermined set points. The primary chamber will be designed to maintain a temperature of 1400°F (600°F in a substoichiometric atmosphere).

3.6.4 SECONDARY COMBUSTION CHAMBER

The secondary combustion chamber will be a refractory lined chamber equipped with an auxiliary burner. The burner will be automatically controlled capable of preheating the chamber and maintaining a predetermined minimum set point temperature during normal operation. The secondary chamber will be sized to provide one-and-a-half seconds gas residence time at a design exit gas temperature of 1800°F. The secondary chamber is insulated and constructed similar to the primary chamber.

3.6.5 MECHANICAL ASH DISPOSAL

The primary combustion chamber will be equipped with a hydraulic ash removal system. The ash removal system will consist of an internal ash transfer ram and a hydraulic slide gate. The internal ash transfer ram will extend into the primary chamber moving the ash along the hearth and discharging it into a 55-gallon drum. As the ash is discharged into the drum, it will be sprayed with water to cool it. This is referred to as a "dry ash system." The drums can then be lifted out, and removed from the ash pit for disposal. The slide gate will provide an air seal to reduce uncontrolled air infiltration into the primary chamber, or fugitive emissions from the primary chamber.

3.6.6 AIR POLLUTION CONTROL SYSTEM

The exhaust gases will be exiting at a temperature of 1800°F and 4000–5000 ACFM. The exhaust will be routed through an insulated exhaust duct that will either pass through the building to the penthouse, or alternately run along the outside of the building to the penthouse. In either case, the exhaust gases will then pass through a heat exchanger to cool them before entering into the air pollution control equipment. A barometric damper will be installed on the exhaust duct providing operational safety in the event the by-pass stack is utilized.

The air pollution control equipment will be a dry scrubber with lime

TABLE 3.9. Capital Cost.

(1) Incinerator	$400,000
(2) Air pollution control system	
(A) dry scrubber	$500,000
(B) wet scrubber	$350,000

injection. The exhaust will be fan-forced exiting through an exhaust stack. The residue from the particulate control device will be brought to the ground floor for disposal.

3.7 CAPITAL AND OPERATING COSTS

Capital and annual operating costs are provided in Tables 3.9 and 3.10, respectively. The figures provided are budgetary, meaning plus or minus fifty percent. The capital costs include equipment, installation and start-up services, but not building construction or stack testing costs. The annual operating costs are based on 8 hours per day, 5 days per week, and 52 weeks per year. The capital cost was annualized, assuming a 15-year equipment life, a 10% interest rate and a capital expenditure of $900,000. All costs are based on a system capable of handling 400 pounds of waste per hour.

A cost savings will be realized since orange and red bag wastes will not have to be removed by BFI. This amounts to approximately $30,000 per year (370 lb/day @ $0.31 per pound).

TABLE 3.10. Annual Operating Cost.

(1) Labor	
2 employees @ $20,000	$40,000
(2) Maintenance and repair	
7% of capital cost	$63,000
(3) Fuel—natural gas	
8.32×10^6 ft^3/yr @ $3.20 @ 10^3 ft^3	$27,000
(4) Electricity	
1.62×10^5 kWH/yr @ $0.1234 per kWH	$20,000
(5) APC chemicals	
9.45 tons/yr @ $400 per ton	$4,000
(6) Ash disposal	
41.6 tons/yr @ $50 per ton	$2,000
Annual operating cost	$156,000
(7) Annualized capital cost	$118,000
Total annual cost	$274,000
Note: Avoided waste disposal cost	
96,200 lbs/yr @ $0.31 per pound	$30,000

3.8 CONCLUSIONS AND RECOMMENDATIONS

C/TA has reviewed the past and anticipated waste profile for the facility. Based upon the waste disposal needs, it appears that a single incinerator operating at 400–500 pounds/hour, at 8–10 hours/day, will handle the waste stream. The concept is for a unit that will be a special (but still stock) unit to handle either pathological or infectious wastes under several operating conditions, different operating hours, and different feed mechanisms.

The proposed system will have an efficient air pollution control system that will be located in the penthouse area of the new building. It is recommended that a dry system (with lime injection) be considered for the air pollution control system. This type of a control system will eliminate a wet, visible plume.

CHAPTER 4

High Efficiency Wet Scrubbers for Particulate and Gas Pollution Control

KENNETH C. SCHIFFTNER, P.E.[1]

4.1 INTRODUCTION

WHEN we discuss the use of modern, high efficiency wet scrubbers in the light of the new Clean Air Act, it is important to understand how the new regulations have come into being.

What is the intent of the revisions? What is expected of me as a user of air pollution technology? What kind of equipment do I need? Is what the government *thinks* I need really best for my company? Is there a common ground? Tables 4.1 and 4.2, relative to particulate and gas control systems, can be used as guides for this.

The Clean Air Act (CAA) revisions imply that the very best technology be applied to emissions source. But what is implied, and what occurs, are sometimes very different. It can be said that today's emissions control technology is driving the law, rather than the law driving the technology. This discussion, along with the rest of the book, may lead you to agree with this assessment.

A good overview of the CAA amendments was presented in *Pollution Engineering* in March, 1992 (Bryant, 1992). Our analysis here will expand upon this document and help you determine the type of equipment to use to meet the CAA requirements.

Some people believe that the technology changes to suit the emissions limits defined by the law. This may be partly true, but it can be argued that in reality it truly is the technology that drives the law.

I would like to provide a little background information and then provide

---

[1]Compliance Systems International, Carlsbad, California.

TABLE 4.1. Particulate Control.

Requirement	Technological Candidates
Low gas volume (under 50,000 ACFM) general particulate	Baghouse, ESP, Venturi scrubber, WESP, proprietary designs
Low gas volume, with heavy metals	Baghouse with additives, spray dryer baghouse or ESP, condensing wet scrubber, condensing scrubber with WESP
Low gas volume, heavy metals, and dioxins	Condensing wet scrubber, condensing scrubber with WESP baghouse and wet scrubber. Condensing scrubber and brinks filter, wet scrubber and WESP
High gas volume (over 50,000 ACFM)	Baghouse, ESP
High gas volume, with heavy metals	Baghouse with additives, spray dryer with reactor and baghouse. Baghouse or ESP followed by wet scrubber
High gas volume, heavy metals and dioxins	ESP with condensing wet scrubber, baghouse with reactor and additives, baghouse followed by wet scrubber and WESP

ESP = electrostatic precipitator.
WESP = wet electrostatic precipitator.

TABLE 4.2. Gas Absorption.

Requirement	Technological Candidates
Low gas volume (under 50,000 ACFM)	Packed towers, spray towers, fluidized bed scrubbers, grid tray towers, mist-type absorbers
Low gas volume, high efficiency	Multiple stages of above
High gas volume (over 50,000 ACFM)	Spray towers, grid towers, fluidized bed scrubbers, spray dryer and baghouse combinations
High gas volume, high efficiency	Hybrid units, units as above in series

Note: For any of the above applications, greater scrubber process control can be expected to meet compliance under the CAA amendments.

a hopefully simplified description of the sizing and application of modern wet scrubbing technology to meet the amended Clean Air Act requirements.

4.2 BACKGROUND

In the years prior to the 1970 Clean Air Act, growing public concern resulted in pressure on government to develop *health risk*-derived air emissions standards. The original Clean Air Act responded to that public need by setting certain emissions standards that were deemed to be protective of the general population's health. Particulate emissions standards based upon production rates, or those that resulted in acceptable ground emissions concentrations were adopted. Some of these requirements were generated through health risk assessments, dispersion models, and other statistical methods to define the levels to which potential pollution sources had to reduce their emissions.

During the 1970s, however, and continuing into the 1980s, a new focus began to appear. The concept of BACT (Best Available Control Technology) appeared in the laws surrounding the CAA. Not only were the health risk code levels part of the "brew", there was now a subjective element (the concept of "best") injected into the requirements.

As an engineer with over 25 years experience in the air pollution control field, the author has a great deal of difficulty with the word "best" when it applies to compliance technology. Perhaps this uneasy feeling is shared by many.

"Best" implies a value judgment based upon comparison of alternatives. Factors such as efficiency, cost, reliability, safety and other items can be combined to define the word "best" as it applies to air pollution control equipment. To a person living next to an incinerator, perhaps "best" means the device that allows the lowest net air emission regardless of the cost. To a worker at the plant, "best" could mean the device that demands the least amount of equipment to maintain. To the accountant for the firm buying the equipment, the word "best" could mean the lowest cost unit that just meets today's codes.

In a compliance framework wherein only the "best" is allowed, energetic vendors can legally seek to exclude all others from the market. If they can get government to use their definition of "best," or, better yet, get the government to directly recommend their type system, the vendor is on easy street. Given the law's definition, you'd seemingly *have* to buy their type system. This is one reason why there was intense lobbying by equipment vendors during the CAA rewrite.

The concept of "best" also contains a time element. "Best" today may be pitifully poor tomorrow. But how do tomorrow's technologies get a chance, if only today's proven best are permitted to be installed?

The point is that the focus has changed from health risk issues, which could to some extent be supported by statistics, to complex value judgments some of which may have nothing to do with a company's long term best interests. The effort may also stifle the development of improved technologies.

A recent article in *Power* magazine (Makanski, 1991) summed up a position many environmental professionals share,

> Above all else, keep this in mind: No one fully understands all the implications of the CAA.

The CAA does not specifically define what type of technology to apply to what emissions source, though some receive free advertising by being mentioned. It does, however, specify achievable reduction efficiencies for certain pollutant groups (such as hazardous air pollutants). To make certain that the reductions were technologically achievable, the CAA writers surveyed certain technologies and used them as the basis of their reduction projections. Basically, the logic sequence was:

(1) To what level can today's "best" system reduce the emissions?
(2) Is the technology proven today, circa 1993 A.D.?
(3) O.K., then do it!

Unfortunately, "number 2" reduced the chances of application of newer, perhaps better technologies that had not been proven at the time of the CAA rewrite.

Given this background, I will try to help you simplify the difficult task of selecting the technology that can help you match the available technology to the realities of the CAA.

4.3 TECHNOLOGICAL CHOICES

In Table 4.3, we list the logic sequence that we recommend in deciding what type of air emissions equipment, if any, you should use. The "if any"

TABLE 4.3. "Logic" Sequence Equipment Selection Scrubbing Technology.

1. Consider process changes
   if not possible, then:
2. Consider system upgrades
   if not possible, then:
3. Consider new equipment
   a. Review current technology
   b. Review emerging technology
   c. Perform a cost analysis
   d. Check compatibility with process

is a key element in your decision. If you do need equipment, you may need more than one type.

I thought at first that I would just describe wet scrubbers. To be fair, however, the modern applications engineer lives in a world of hybrid technologies. Wet scrubbers are used successfully coupled with thermal oxidizers to remove acid gases, following baghouses for condensation scrubbing of submicron residuals, and before or after wet electrostatic precipitators. Figure 4.1 shows a diagram of a modern hybrid type system. In this one, a dry/wet scrubber is followed by a condensing wet scrubber for dioxin and heavy metals control. Given the use of hybrid technologies, we will try to describe the many variations you may find.

Following our logic sequence, we will first suggest and describe process changes and equipment upgrades that you should consider *before* considering any air pollution control device.

After that, to simplify things, we will separate our discussion into two subsets, particulate control and gaseous emissions control, and within those groups, focus on high volume sources and on low volume sources. High volume sources will be defined as those devices handling gas volumes of over approximately 50,000 scfm. Low volume sources will be those sources at or below 50,000 scfm. The reason for this separation by volume is that wet scrubbers (with a few exceptions) increase in cost of operation per cfm treated dramatically after certain size ranges given their pressure drop requirements.

4.4 PROCESS CHANGES

Many processes currently in use in the United States were designed without serious concern for environmental issues. One of the greatest sources of emissions reductions comes from process changes, but these can be very difficult to accomplish "politically" in a company. The production engineering group may be totally different from the environmental group, with the latter having no authority over the conduct of the former.

Management should recognize, however, that one of the lowest cost methods of meeting compliance can be to simply modify the process.

Medical waste incineration is a good example. Hospitals used to burn nearly all of their waste in their "medical waste" incinerators. Plastics, paper, batteries, film, etc., were burned along with infectious waste. Items were purchased in brightly colored coated boxes (whose pigments contained heavy metals), batteries were used and thrown into the dumpster (thus placing mercury or cadmium in the waste disposal stream), and chlorinated plastics were used which upon combustion yielded hydrochloric acid fumes that had to be removed from the air.

By properly purchasing items, and recycling or segregating problematic

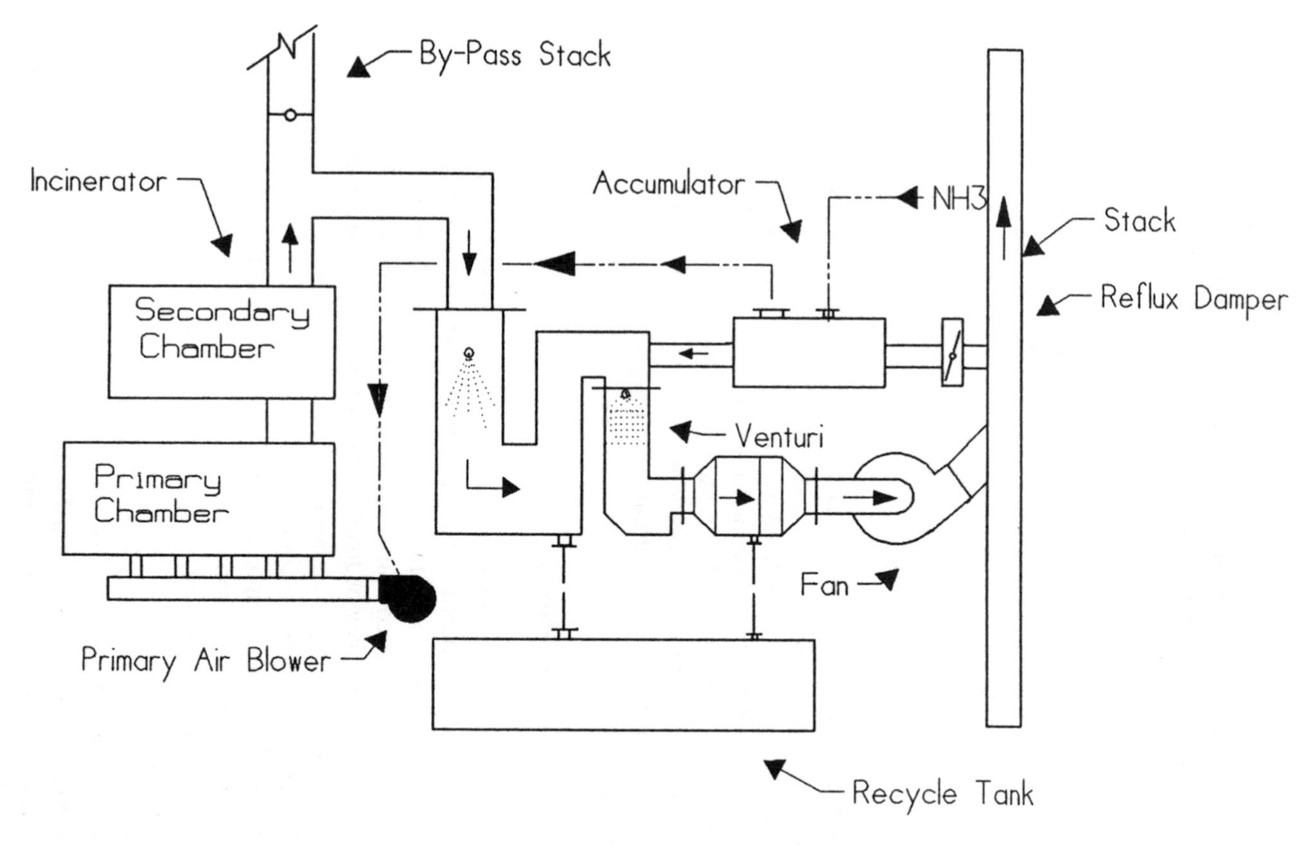

Figure 4.1 Integrated incinerator/scrubber system.

waste, the potential emissions can be reduced with no extra air pollution control expense. And, the reductions that are made can reduce the cost of any equipment that is required.

In painting operations, many firms have switched to water soluble paints (to reduce VOC emissions), electrostatic spraying techniques (to reduce overspray losses), and have installed more carefully controlled paint curing operations.

Process changes, though sometimes painful, are a very powerful way to match technology to meeting the requirements of the Clean Air Act.

4.5 UPGRADING

If your initial analysis suggests that only minor improvements in removal are needed, consider an upgrade. Most air pollution control companies have engineers who can suggest methods of improvement of their equipment. Consultants also supply this service and may be more capable of suggesting Hybrid systems such as shown in Figure 4.2.

Figure 4.3 shows a modified Venturi scrubber wherein the old throat is replaced by a new throat module. In this design, the old "damper blade" type throat mechanism would be removed and the new module would be inserted in its place. A spray header is added above the throat module to disperse scrubbing liquid. In this design (patent applied for), the throat widths are drastically reduced to minimize gas sneak-by that can occur in larger throats.

Sulfur dioxide absorbers can sometimes be improved through the addition of buffering agents (adipic acid) in the recycle liquid, forced oxidation of the recycled scrubbing liquid, or the addition of contact grids or extra spray levels in the absorber vessel. On systems running frequently in a "turn down" mode, flue gas recycle designs can send previously scrubbed gases back through the tower for an additional separation pass, thus lowering total emissions.

Venturi scrubbers can usually be improved by adding pressure drop. Though this often requires a new fan and a modified Venturi throat, the cost can be lower than a complete scrubber replacement. Droplet elimination may also have to be improved if the higher pressure drop Venturi generates greater quantities of smaller droplets. The scrubber recycle water chemistry can also be changed to foster greater particulate or acid gas removal.

4.6 NEW EQUIPMENT

Assuming that you cannot make sufficient process changes and you cannot upgrade enough to meet the emissions reduction target, you may have to purchase new equipment.

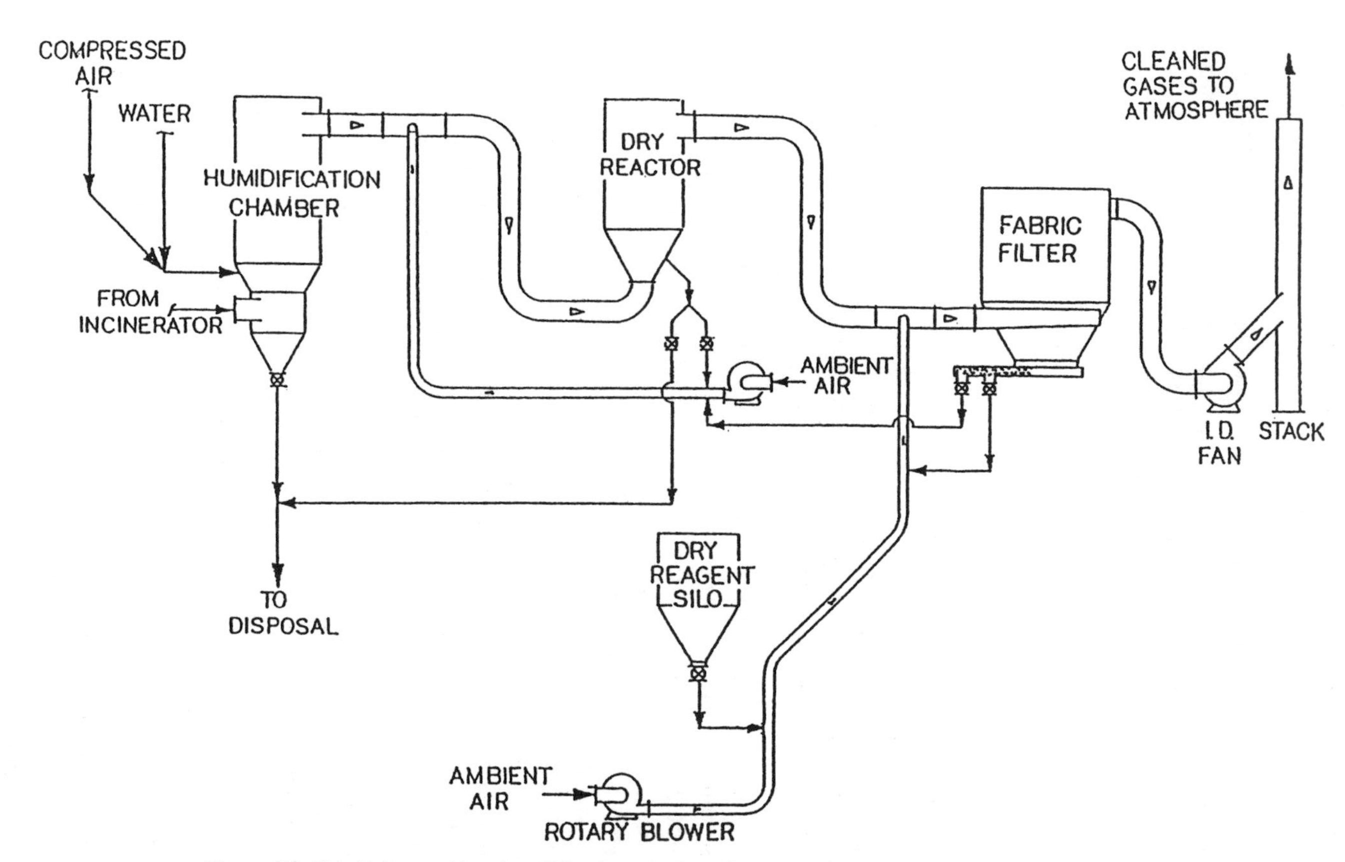

Figure 4.2 Hybrid dry scrubber: humidification, reaction, filtration technology are used in one system.

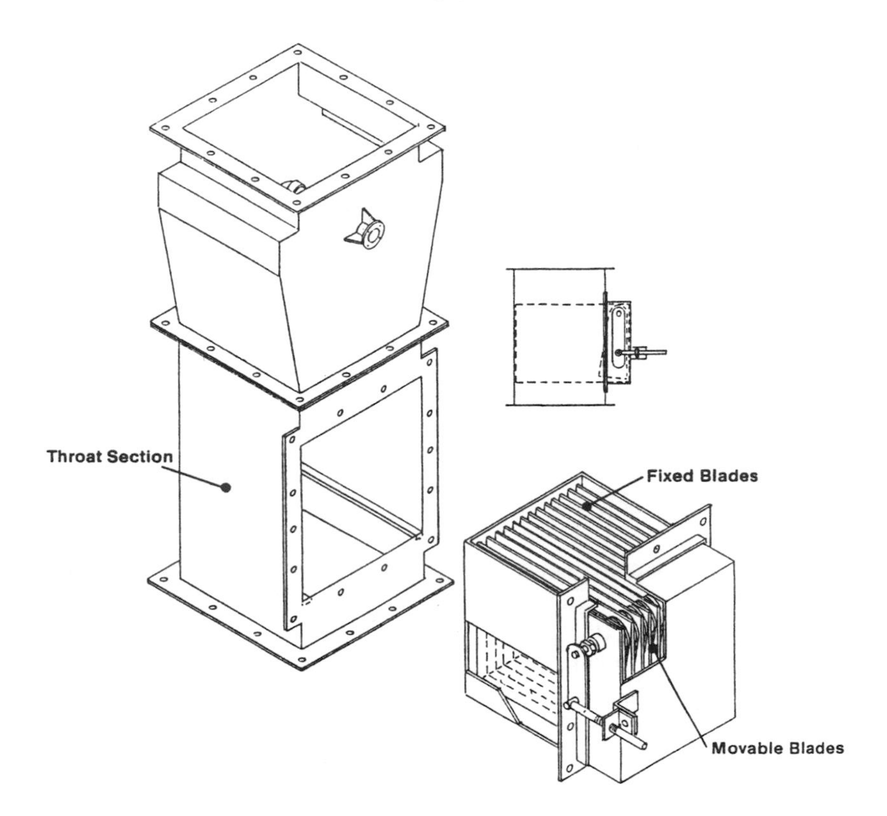

Figure 4.3 Narrow gap Venturi scrubber (NGV).

In general, the CAA has reflected the success of "hybrid" type air pollution control equipment. "Hybrid" means that a variety of technologies are used to reduce the emissions, not just one. Compound problems require compound solutions (see Figure 4.2).

Though this chapter focuses on Wet Scrubbers, in the interest of completeness, hybrid type systems should be mentioned.

The modern age of gas cleaning technology has resulted in systems wherein a baghouse is followed by a condensing wet scrubber which is, in turn, followed by a wet electrostatic precipitator. Systems are not uncommon in Europe where spray dryers are followed by baghouses, which are then followed by activated carbon systems for dioxin and heavy metals separation.

In matching technology to meet the CAA requirements, expect to have your resulting system incorporate a variety of technologies in series, each one performing a specific task.

We will try to summarize the application of modern technology for some of the key elements of the new Clean Air Act.

4.7 VOC CONTROL

VOC (volatile organic carbon) compounds may be water soluble, water insoluble, condensible, and non-condensible.

If the gaseous emissions are *insoluble in water,* are in great enough concentration to provide some heat of combustion, and are not worth recovering, thermal oxidation (Kottke, 1992) is a possible solution. These techniques include:

(1) Direct flame incineration
(2) Catalytic incineration (assuming that the emission is free of catalyst-poisoning metals)
(3) Regenerative thermal oxidation
(4) Recuperative thermal oxidation (A heat exchanger is used to increase thermal efficiency.)

Any of these incinerators may have to be equipped with acid gas scrubbers if the VOC being destroyed is halogenated. The scrubbers used are packed towers, spray towers, tray towers, and fluidized contact type units. They operate in vertical (counterflow) modes at gas rates of 3 – 20 feet per second gas velocity and in crossflow modes of up to 10 feet per second. They typically recirculate pH-adjusted scrubbing liquid at the rate of 3 – 20 gallons per 1000 actual cubic feet of air treated per minute.

If the gaseous emissions are *insoluble in water*, are low in concentration (below about 1000 ppmv), and may be worth recovering, activated carbon or zeolite adsorption systems should be investigated. These devices are available in "throw-away" designs or regenerative designs. Each act as VOC accumulators. The carbon type units are typically purged (stripped) with steam, or nitrogen, and the zeolite units with hot air or other heated carrier gas. The stripped compounds are destroyed or recovered. Condensation systems may also be used if the vapor contains recoverable VOCs. A cost analysis should be conducted on individual units.

If the gases are predominantly *soluble in water* with some insoluble components, you may want to consider water or emulsion scrubbing. If the concentrations are low and the gas volumes are too high to economically thermally oxidize the material, scrubbing with cold water or emulsions containing oil and a coupling agent may prove satisfactory. These towers operate at about 5 – 7 feet/second vertical velocity if in a counterflow mode. With emulsion scrubbers, the emulsion can be broken, the oil separated and treated (some of these combined systems are patented). Water scrubbers with peroxidation type (ultraviolet light destruction of waterborne VOCs) water treatment systems may find favor on smaller systems. Pilot testing is highly suggested.

If the VOCs are in the form of an *aerosol*, consider fiberbed coalescing technology. In these devices, Brownian motion is used to trap aerosols on densely packed fibers, allow them to accumulate, and drain.

An emerging alternative VOC and odor control technology is biofiltration. Used extensively in Europe, there are now successful systems operating in the U.S. These units are like large biological packed adsorber/absorber towers, wherein the packing media is comprised of wood chips and supporting organic material to reduce the bulk density. Gas velocities are usually far below 3–4 feet/second and contact beds are typically 2–3 feet deep minimum. Biofiltration is not an exact science, but can allow cost effective VOC control where the target VOC has little value, the inlet concentrations are consistently low, and the tower can be kept warm (above 80° F, but below 120° F). Some other variations consist of imbedded pipes in soil.

4.8 PARTICULATE REMOVAL

If the gas stream is a high volume (above 50,000 ACFM) source, usually baghouses and electrostatic precipitators (ESP) are favored given their lower pressure drops when compared to alternative technologies. These devices clean gas streams at pressure drops below 10″w.c. Precipitators and baghouses typically have "can" velocities (i.e., the plug gas flow rate through the housing) of 2–8 feet/second. Table 4.1 shows a summary of the technological candidates that could be used to control particulate under the CAA amendments.

Don't expect to ONLY use a baghouse or an ESP under the Clean Air Act amendments, however. If the emission contains volatile heavy metals such as mercury, arsenic, lead and cadmium, you may need to use additional technologies. A hazardous waste incinerator at DuPont, in Orange, Texas, for example has a baghouse followed by a condensing wet scrubber that includes a collision-type Venturi scrubber. The latter system is intended to remove condensible heavy metals and phosphorous pentoxide fume. Though various baghouse vendors are marketing various additive (coke, activated carbon, silica) injection systems to enhance dry collector performance, the thermodynamic properties of these more volatile heavy metals lend themselves to effective removal by wet systems that utilize cooling and condensation techniques.

Systems have been proposed that include a baghouse as a primary collector of particulate, a condensing wet scrubber for volatile heavy metals and dioxin control, all followed by a wet ESP for residual particulate removal (particles below 0.3 microns). There is no carbon or coke injection at all, and the waste volume is minimal.

Needled felt baghouse media (such as Huyglas 1701, Air Purator Corp.), polyamide material (P-84, Lenzing Corp.), and membrane material (Goretex™, W.L. Gore Co.) are providing high particulate removal rates.

For low gas volumes (below about 50,000 ACFM), the selection problem becomes more difficult since a wide variety of gas cleaning devices can be used. In general, the cost per cfm treated using a precipitator increases as the volume decreases because items such as power supplies and controls do not have proportionally lower costs.

For low gas volumes, modern Venturi scrubbers can be used. Wet scrubbers remove more total mass of pollutants on medical waste incinerator applications than dry scrubbers. One type in operation at hospital waste incinerators in New York and Florida uses condensation scrubbing and a high efficiency Venturi scrubber. It produces outlet particulate loadings of well under 0.010 grs/dscf corrected to 7% oxygen and removes over 99.9% of the HCl. Other technologies currently applied to low gas volumes are a variety of proprietary wet scrubbers (rotary atomizer types, countercurrent spray type, and eductor type) that have also proved successful.

Venturi scrubbers typically use inlet gas velocities of 50–60 feet/second and throat velocities of 100–400 feet per second. Condenser/absorbers operate at vertical velocities of 3–7 feet per second. A unique Venturi, developed by Calvert Environmental and called the Collision Scrubber, is

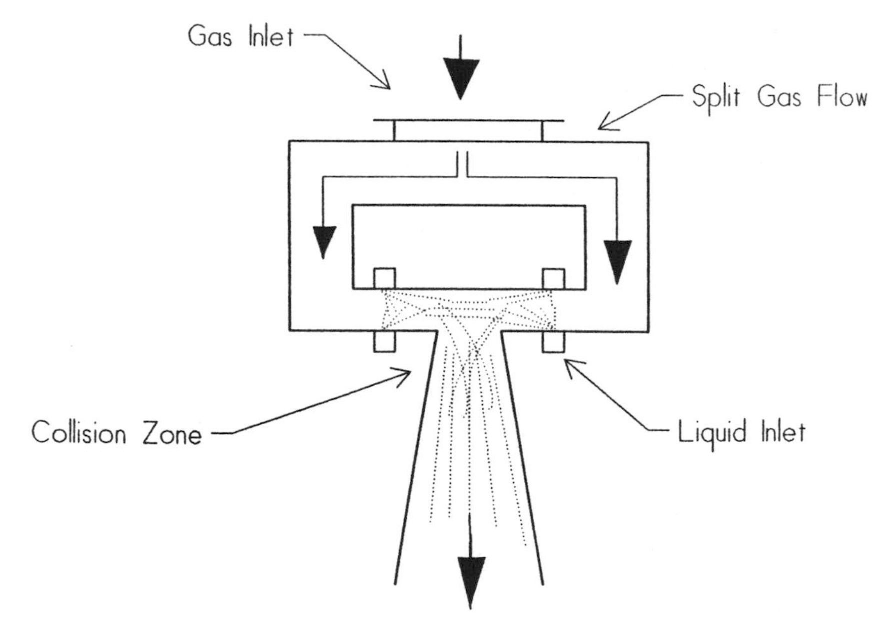

Figure 4.4 Collision scrubber.

shown in Figure 4.4. Another unique scrubber is shown in Figure 4.5. This is the Catenary Grid Scrubber™ supplied by Compliance Systems International. Both are extremely effective for particulate removal at lower than expected pressure drops. Note that Figure 4.5 also shows one method for draft control using flue gas recycle through a scrubber.

If your source contains over 15% submicron particulate, and you must meet code requirements of under 0.015 grs/dscf, consider the use of the condensation scrubber. These devices have been used for years on hazardous waste incinerators and are now finding favor on medical waste incinerators where residual submicron salts are present. These devices first saturate the gas stream with water vapor, then condense the vapor in the gas stream by direct contact with cold water, then use a Venturi scrubber for particle collection. As mentioned earlier, they have proven an ability to scrub to under 0.01 grs/dscf while removing 99.9+% of acid gases, far better performance than nearly any other system. And they also reduce the stack plume.

If the emission must be reduced even further, wet scrubbers are now being followed by fiberbed type "candles." These devices allow extremely low outlet loadings and work best where the captured particulate can be dissolved in cleaning liquid. Some of these fiberbeds are being made of combustible material and can be incinerated for disposal. Units are in operation on nerve gas incinerators and on other hazardous waste units. Applicable ranges for effective particle collection by several mechanisms are depicted in Figure 4.6 in generalized form.

4.9 GASEOUS EMISSIONS CONTROL

When using wet scrubbers for gaseous emissions control, the predominant technique used is gas absorption. All wet gas absorbers work by extending the surface area of a liquid and mixing it with the contaminant gas. Some do this by using sprays, others use packing, others use grids or trays. Table 4.3 shows several gas absorption options.

If the application involves high gas volumes, look towards spray towers or grid type counterflow absorbers. These devices can be designed to handle volumes of over 100,000 ACFM in a single vessel. Expect to have to treat the scrubbing slurry to enhance performance. Expect to have a more exotic control system to automatically monitor the system. These towers operate at vertical velocities of 6–20 feet/second, with 6–8 feet/second used for spray towers and 16–20 feet/second used for the Catenary Grid Scrubber™ type tower. Spray towers typically recirculate 20–100 gallons of liquid per 1000 ACFM of gas and grid towers use about 10–30 gallons/1000.

If the treated gas volume is under the low volume classification, packed

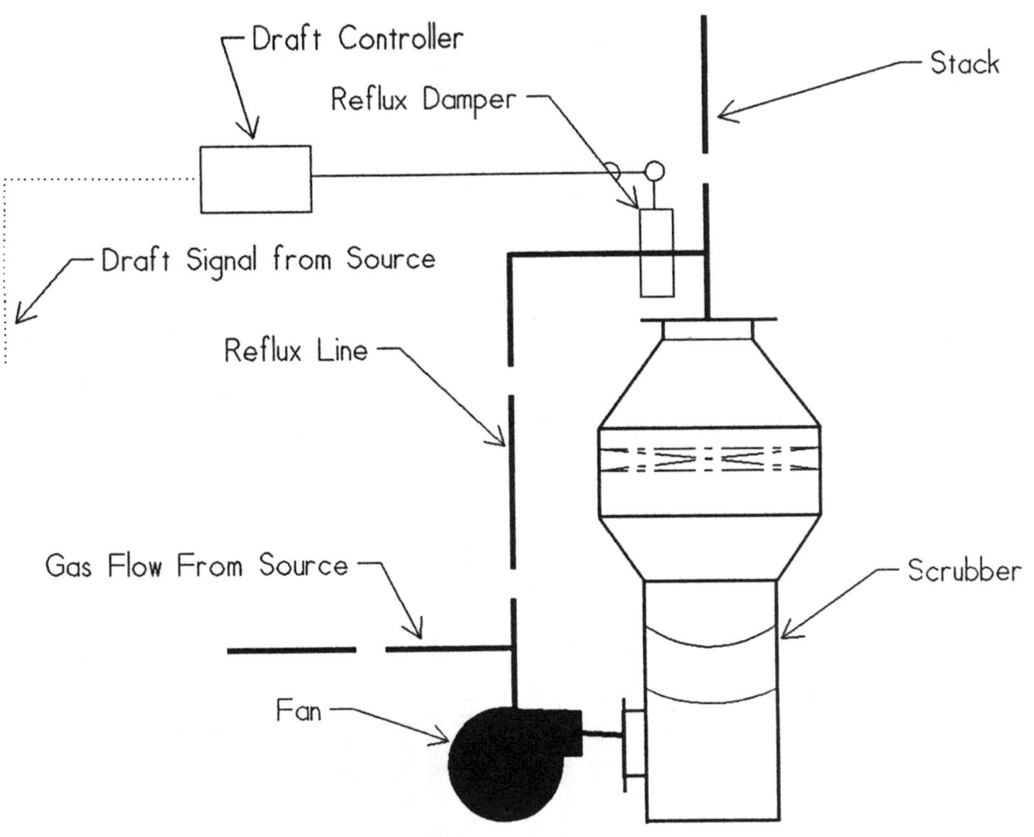

Figure 4.5 Draft control using flue gas recycle.

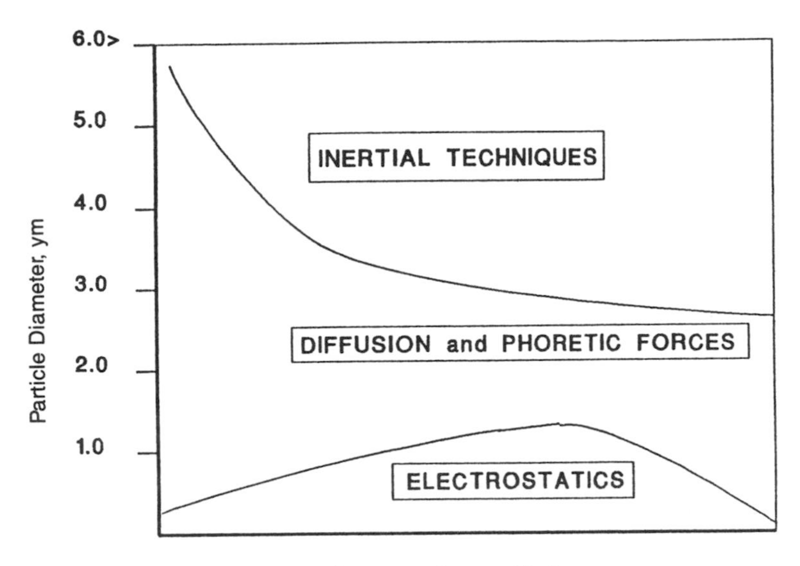

Figure 4.6 Capture technique range.

tower absorbers using dumped-type or structured packing media still find favor. If you have an existing tower, it can sometimes be salvaged either as the first stage of a two (2) stage system and/or by changing the packing type. Various packing vendors now will size the tower for you if you provide the known inlet concentrations, temperatures, flows and pressures, and the target outlet conditions.

4.10 OTHER FACTORS

An often overlooked method of wet scrubber improvement is better control of the system. The CAA amendments force you to take better control of your gas cleaning system. This is often not difficult and can actually save you money.

For example, increasing numbers of controlled air hospital waste incinerators are being fitted with flue gas recycle systems for draft control (Schifftner, 1991). (See Figure 4.5.) The scrubber throat remains fixed. Wet scrubbers are sometimes criticized for "creating a water pollution problem from an air pollution problem." With modern process control, however, they are unique in that they can control and stabilize the pollutant *at the source*. Many competing devices simply push the problem down the line to the landfill. Wet scrubbers exist today where heavy metals separation and recovery can occur right at the scrubber location so that the waste ultimately sent to the landfill is less toxic, more stable, and of lower quantity than other technologies. Wet scrubbers can remove dioxins from the air, and destroy them at the site. Few competing systems can make that claim.

I trust that the overview has helped you understand how to select wet gas cleaning technology to solve your air emissions problem. The message is to think about process modifications or upgrades first, then, if you have to, think about the technology you will need to reach compliance. Don't be surprised that you may need hybrid technology. In today's world, the question isn't black or white, or dry or wet; it is what will work for me?

4.11 REFERENCES

Bryant, C., "EPA Releases Draft Hazardous Organic NESHAP," *Pollution Engineering*, March 15, 1992, page 25.

Makansi, J., "Clean Air Act Amendments: The Engineering Response," *POWER Magazine*, June 1991, page 14.

Kottke, L.R., "VOC Technology Catches Up to Regs," *Environmental Protection*, March 1992, pp. 22–27.

Schifftner, K.C., "Integrating Incinerator, Scrubber and Waste Disposal: The Key to Successful Medical Waste Control," Paper 91–30.2, Air and Waste Management Association, Vancouver, B.C., June 1991.

CHAPTER 5

The Control of Gaseous Pollutants Through Adsorption

FRANK L. CROSS, JR., P.E.[1]

5.1 INTRODUCTION

ADSORPTION is a physical process by which gaseous molecules are collected on the surface of a solid. Adsorption is characterized in Tables 5.1 and 5.2. Activated carbon is the most widely used adsorbent, although other materials such as silica gel, alumina, and bauxite may be used as adsorbents, as noted in Table 5.3. Gases, liquids or solids can be selectively captured and removed from air streams by the adsorption process. The adsorption mechanism is complex. Usually the adsorption rate is small unless the solid adsorbent is highly porous and possesses fine capillaries. The most important characteristics of adsorbents are their large surface-to-volume ratios and preferential affinity for individual components.

There are several theories to explain the selective adsorption of certain vapors and gases. Adsorption may be due to chemical combination of the gas with the free valences of atoms on the surface of the solid in the molecular layer as was proposed by Langmuir in 1916. Another theory is that the adsorbents exert strong and attractive forces so that many adsorbed layers are formed. These layers are under pressure due to the layers on top and the attractive force of the surface of the adsorbent. In other adsorption cases, the evidence indicates that adsorption is caused by the liquification of the gas and its subsequent retention by capillary action in the very fine pores of the adsorbent. Each of the three theories is supported by some evidence. In many cases, the adsorption phenomena are probably due to a combination of the three mechanisms. Adsorption with activated charcoal is due mainly to molecular capillary condensation.

[1]President, Cross/Tessitore & Associates, P.A., Orlando, Florida.

TABLE 5.1. **Adsorption.**

90% + efficiency
Handles high VOC concentrations
Humidity > 45% decreases efficiency
Design problems with VOC mixtures
Higher MW compounds adsorbed mroe readily

TABLE 5.2. **Adsorbent Characteristics.**

1. Large adsorption capacity
   a. Large specific area
   b. Be "activated"
2. Be a specific adsorbent to meet the problem
3. Granular in nature to prevent high head losses through the bed
4. Be resistant to breakage
5. Be easily regenerated
6. Be as inexpensive as possible

TABLE 5.3. **Adsorption Materials.**

Adsorbents can be divided into four classifications:

1. Charcoal and activated charcoal
   a. Hydrocarbons
   b. Nonpolar gases
2. Hydrous oxides (silica gel, aluminum oxide, magnesium silicate)—polar gases and vapors
3. Finely divided metals—generally used as catalysts
4. Natural materials (Fuller's earth, clays, bauxite)—low adsorptive capacity

In air pollution control engineering, it really does not matter why adsorption occurs; adsorption does work and can be used as an air pollution control operation.

Activated carbon and silica gel, alumina, and bauxite are used for selectively adsorbing certain gaseous constituents from gas streams. In air pollution control, activated carbon has the widest application as an adsorbent. (Alumina and bauxite are used chiefly for dehydration.) Activated carbon adsorbs organic gases and vapors selectively from the air stream even when water is present. Silica gel will also adsorb organic and inorganic gases; however, in the presence of water vapor, silica gel selectively adsorbs water vapor. Table 5.4 lists various adsorbent recovery rates. A process exhaust stream is passed through a fixed bed or adsorbent prior to being discharged to the atmosphere.

Adsorption of the organic vapor occurs in two stages. Initially, adsorption is rapid and complete, but eventually the carbon continues to remove material at a sharply decreasing rate. After a period of time, the adsorbent becomes saturated and traces of solvents begin to appear in the exhaust stream. This is the breakpoint. When this breakpoint occurs, the adsorbent material must be taken off stream and replaced.

The adsorption beds that have been saturated are usually regenerated for re-use. These regenerative systems incorporate reactivation of the adsorbent by desorption and recovery of the adsorbed material for re-use or disposal. Adsorption systems may also be non-regenerative, i.e., replacement of the spent adsorbent with fresh material is utilized. However, for other than the small systems, non-regenerative systems are economically unfeasible. Regenerative systems usually employ two or three carbon beds for adsorption. In a 2-bed system, one bed is used while the other is being reactivated. In a 3-bed system, one bed is adsorbing while one bed is being reactivated, and the other bed is being cooled prior to adsorption. In any case, after the solvent is stripped from the carbon bed in the reactivation step, the carbon must be cooled and dried.

5.2 SOLVENT RECOVERY

In specifying an adsorption system for solvent recovery, a series of basic factors must be considered. A typical data sheet for equipment specification is shown in Tables 5.5 and 5.6. Adsorbent applications are noted in Table 5.7. Of course, the polluted gas stream flow rate must be established. The temperature of the effluent is important since 120°F is the maximum temperature for operation of a carbon adsorption bed. Cooling equipment must be provided if the gas stream temperature is above 120°F. Impurities such as particulates must be removed prior to the carbon bed since particu-

TABLE 5.4. Approximate Solvent Recovery Capacity (lb/hr per bed).

CFM at Less Than 100°F										
One Bed	Two Beds	Acetone	Chloroform	Ethylene Dichloride	Freon	Heptane	Hexane	Methylene Chloride	Trichlor-ethylene	Perchloro-ethylene
700	—	8	10	15	8	6	6	10	15	20
750	—	28	35	52	28	21	21	35	52	70
750	1,300	28	35	52	28	21	21	35	52	70
800	1,400	42	50	78	42	18	18	50	78	100
700	—	12	15	23	12	9	9	15	23	30
800	—	28	35	56	28	21	21	35	56	70
800	1,350	28	35	56	28	21	21	35	56	70
1,400	—	32	40	60	32	24	24	40	60	80
1,700	3,000	80	100	150	80	60	60	100	150	200
3,000	5,500	120	150	225	120	90	90	150	225	300
3,800	7,500	240	300	450	240	180	180	300	450	600
5,000	10,000	360	450	675	360	270	270	450	675	900

Other solvents may also be recovered.

TABLE 5.5. **Adsorption Systems.**

Listed below are the major factors affecting adsorption system efficiency:

1. Surface area of the sorbent
2. Nature of adsorbent (it is specific to the absorbent)
3. Temperature
4. Gas velocity through the unit
5. Concentration of the adsorbate
6. Relative saturation of adsorbent

TABLE 5.6. **Factors in Adsorption System Design.**

1. Contact time between adsorbent and adsorbate
2. Total adsorptive capacity of the adsorbent
3. Uniform distribution of air flow over the surface of the adsorbent
4. Resistance to air flow
5. Provisions for regeneration/methods and time required
6. Quantity of fluid handled per unit time (i.e., ft^3/min)
7. Amount of material to be adsorbed

TABLE 5.7 **Applications.**

Coating operations
Degreasing operations
Gasoline marketing
Odor control

TABLE 5.8. Lower Explosive Limit.

	LEL		
	Percent	ppm	25% LEL
Methyl ethyl ketone (MEK)	1.8	18,400	4,600
Butyl acetate	1.7	17,300	4,320
Cellosolve	2.6	26,700	6,670
Cellosolve acetate	1.7	17,400	4,350
Napththa	0.92 – 1.1	9,290	2,320
Toluene	1.27	12,600	3,150
Xylene	1.0	10,100	2,520
Mineral spirits	0.77	7,760	1,940

lates will foul the bed, will quickly reduce and negate the adsorption properties of the carbon, and will render replacement of the beds necessary.

The concentration of solvents in the effluent must be maintained below 25 percent of the lower explosive limit to avoid possibilities of explosion and combustion hazards. Insurance companies will insist upon this. Typical values of lower explosive limits are given in Table 5.8.

5.3 APPLICATION

Generally, solvent recovery is economically advantageous when the concentration of solvent in the effluent stream is above 1,000 ppm. To comply with air pollution control regulations, solvent or organic vapors may have to be collected when the concentration is below 1,000 ppm.

Before the adsorption system is specified and surely before the system is ordered, it must be decided *where* the system will be installed. Absorption systems for solvent recovery on an industrial scale are not small. Almost any configuration can be engineered, but space is an important consideration.

An adsorption system for solvent recovery consists of super-heater, condenser, decanter, blower, blower motor, cooling tower, water pump, filters, filter housing, carbon charge, carbon vessels, screens, ducts, and piping.

5.4 EXAMPLE SYSTEM SPECIFICATIONS

A summary of parametric and calculations is given as Table 5.9. The amount of toluene removed at 90% removal efficiency is obtained using the

TABLE 5.9. Summary of Adsorption Parameters and Example Calculations.

Parameter	Design Range	Data for Example	Remarks
VOC BP, °C	20 – 115	110.6	Good
VOC MW, lb/lb mole	50 – 200	92.1	Good
LEL, %	10 – 50	10%	Good
VOC, Concentration range, ppm	500 – 5000	1400	Would be lower for odor control
Gas throughput velocity, ft/min	80 – 100	100	100 max for 95% efficiency
Working capacity, % of saturation capacity	25 – 30	25	May be determined directly from test data
Transfer zone depth, inch	6 – 18	18	For 90% efficiency use deep bed
Bed depth limit, ft	4 max	1 1/2	For fixed horizontal beds
Pressure drop, inch/water foot bed depth	3 – 15	6	See manufacturer's data
Carbon density, lb/ft^3	27 – 30	30	
Cycle time, minutes	60	60	
Regeneration steam @15 psig, lb/lb carbon	0.25 – 0.35	0.3	Used only during regeneration
Condenser water, gpm/100 lb steam	12	12	May be cooled and recycled; used only during regeneration
System size, scfm		20,000	
Toluene recovered, lb/hr		309	
No. of units in parallel		2	
Activated carbon, lb total		14,580	
Steam @15 psig, lb/hr		1,460	
Condenser water, gpm per hour		175	
Total pressure drop, inches water		20	Varies with ducts, etc.
Fan energy, hp		120	Varies with ducts, etc.

gas flow rate in scfm, the molecular weight of toluene (92.1 lb/lb mole) and the ideal gas relationship of 386 scf/lb mole:

$$(20{,}000\ \text{scf})\ \text{gas}\ \frac{(0.0014\ \text{ft}^3\ \text{toluene})}{(\text{ft}^3)}\ (0.90)\ \frac{(92.1\ \text{lb/lb} - \text{mole})}{(386\ \text{scf/lb mole})} =$$

$$6.01\ \text{lb/min of toluene removed}$$

The amount of this type of carbon required in the transfer zone based on a 1-hour cycle is:

$$\frac{(6.01\ \text{lb toluene})}{(\text{min})}\ \frac{(60\ \text{min})}{(\text{hr})}\ (1\ \text{hr})\ \frac{(\text{lb carbon})}{(0.075\ \text{lb toluene})} = 4810\ \text{lb carbon}$$

At a density of 30 lb/ft^3, the transfer zone volume is:

$$(4810)\ \frac{(\text{ft}^3)}{(38\ \text{lb})} = 160\ \text{ft}^3$$

Bed cross section area is found using the velocity of 100 ft/min maximum and the acfm:

$$\text{acfm} = (\text{scfm})\ \frac{(460 + t)}{(530)}\ \frac{(760)}{(P)}$$

Where t is inlet gas temperature in °F and P is gas pressure in mm Hg:

$$\text{area} = \frac{20{,}000\ \text{acfm}}{100\ \text{ft/min}} = 200\ \text{ft}^2$$

For this adsorbent, assume the strength is capable of supporting a bed depth of up to four feet maximum. This amount of carbon in a 4-foot bed would be:

$$200 \times 4 = 800\ \text{ft}^3$$

$$(800)\ (3) = 24{,}000\ \text{lb max.}$$

The required bed transfer zone depth is only:

$$\frac{(150\ \text{ft}^3)}{(200\ \text{ft}^2)}\ \frac{(12\ \text{in})}{(\text{ft})} = 9.6\ \text{inches}$$

TABLE 5.10. **Effects of Problems with Adsorption Units.**

Problem	Operation	Emissions
Highly exothermic solvents (e.g., ketones, phenols) present in inlet VOC stream.	Hot spots, bed fires may occur resulting in adsorption capacity loss.	Exhaust VOC concentrations in excess of emission limits.
Condensibles transfer pump leaking.	Leaking condensed VOCs.	Vaporization of leaked VOCs.
Control system poorly maintained.	Poorly controlled system.	Potential excess concentration of VOC in adsorber exhaust stream.
Plugged prefilter.	Excessive pressure drop in vapor line.	Increase in bed exhaust concentrations.
Blower (fan) failure (e.g., bearings, belt, motor).	Reduced flow to adsorber.	Increase in bed exhaust concentrations.
Inlet VOC stream relative humidity >50 percent.	Water vapor competes with VOC for adsorption sites.	Breakthrough occurs much sooner.
Nonregenerable compounds present in inlet VOC stream.	Bed fouling, less adsorption capacity.	Breakthrough occurs much sooner.
Corrosion in adsorber.	Pressure loss in bed.	Fugitive emissions from adsorber vessel.
Vapor stream inlet temperature higher than design value.	Revaporization of low boiling compounds will occur.	Exhaust VOC concentratiosn higher than design value.
Inlet vapor stream relative humidity <20 percent	Potential excess heat buildup in bed.	Breakthrough occurs much sooner.
Gradual loss of adsorbent due to entrainment.	Loss of adsorption capacity, premature saturation of bed.	Breakthrough (emission of uncaptured VOC) occurs much sooner.
VOC inlet vapor concentration higher than design value.	Adsorbent is prematurely saturated. Also, possible hot spots and bed fires may occur.	Premature breakthrough.
Emergency bypass damper opened.	Loss of stream to adsorber.	Emissions bypass control unit and are vented to atmosphere.
Regeneration steam traps not bled.	Steam becomes saturated (wet).	Bed is saturated with water and cannot adsorb VOC.
Adsorbent not inspected, replaced on regular schedule.	Buildup of particulates, non-regenerable organics on adsorbent may occur.	Premature breakthrough.
Internal vessel liner chipped, abraded.	Eventual corrosion, erosion of vessel walls.	Fugitive emissions exit through vessel walls.

This depth does not give the 18″ minimum and is not adequate.

In order to meet the design requirements of velocity ≤ 100 ft/min, a transfer zone of 18″ minimum, and sufficient carbon to satisfy working capacity requirements, the following system could be specified for this example.

Use three vessels in parallel with about 100 ft^2 cross section area in each. Two vessels could adsorb while one is desorbed, dried, and on standby.

Cylindrical vessels about 6 feet in diameter by 18 feet long with 18 inches, or deeper, beds could be used.

Amount of activated carbon in each vessel is:

$$(6 \text{ ft})\,(18 \text{ ft})\,(1.5 \text{ ft})\,\frac{(30 \text{ lb})}{(\text{ft}^3)} = 4860 \text{ lb}$$

Total carbon is:

$$(3)\,(4860) = 14{,}580 \text{ lb}$$

Low pressure steam required for desorption in 1-hour cycles is:

$$\frac{(0.3 \text{ lb steam})}{(\text{lb carbon})}\,(4860 \text{ lb}) = 1460 \text{ lb/hr}$$

Condenser water is:

$$\frac{(12 \text{ gpm})}{(100 \text{ lb steam})}\,(1460 \text{ lb/hr}) = 175 \text{ gpm rate for each hr}$$

Assuming ducts and valves add another 11 inches to the adsorber 9 inches of pressure drop, the total pressure drop would be 20 inches. Therefore:

$$\text{Fan HP} = (3 \times 10^{-4})\,(20)\,(20{,}000) = 120$$

5.5 EQUIPMENT COMPARISON

For carbon adsorption systems, one usually finds

- high capital cost
- only simple solvents reclaimed
- medium to large systems requiring constant operating personnel
- desorbed solvents being destroyed
- high boilers in the emissions speeding carbon degeneration
- complicated equipment required – stripping towers, boilers, decanters
- filtration required to protect carbon from particulate and/or dirt as an ongoing cost

- potential fire hazard due to spontaneous combustion
- an acceptable disposal procedure needed for the carbon

5.6 OPERATIONAL PROBLEMS

There are a number of operating problems which can result in substantial decreases in the VOC control efficiency. These problems include the following:

- an increase in the inlet gas temperature and/or the adsorber vessel operating temperature
- an increase in the feed rate VOC vapor to the adsorber vessels
- loss of carbon adsorption activity
- deterioration of the physical condition of the carbon bed
- increase in inlet stream moisture content
- incomplete capture of the VOC emissions from the process sources

Effects of problems are listed in Table 5.10.

CHAPTER 6

Preparing Fabric Filter Specifications

FRANK L. CROSS, JR., P.E.[1]

6.1 INTRODUCTION

THIS chapter presents procedures for preparing specifications, using fabric filters as the example. Fabric filtration systems range in size from the smallest in-plant dust control system exhausting a few hundred cubic feet per minute to mammoth air pollution control systems on the most sophisticated industrial processes. Obviously there will be a qualitative as well as quantitative difference in the information conveyed by buyer to seller in the former case as opposed to the latter. For purposes of this chapter, regardless of the size or complexity of the system, the information conveyed by the buyer to the seller will be called *specifications*.

Whether simple or complex, developed in-house or by outside A&E firms, the specifications for a fabric filtration system should accomplish five basic objectives:

(1) Inform the manufacturers as to system design parameters
(2) Limit the scope of the seller's performance
(3) Inform the manufacturers what the buyer wishes to buy in terms of
    (a) System configuration
    (b) Details of construction of system components
(4) Permit the buyer to make an intercomparison of the various bids on an "apples-to-apples" basis
(5) Establish a starting point for negotiations on terms and conditions of the ultimate sales contract

[1]President, Cross/Tessitore & Associates, P.A., Orlando, Florida.

It is assumed that the reader is familiar with the fundamentals of fabric filtration and with the generic types of fabric filters, that is, shaker, pulse jet, etc. If the reader is unfamiliar with fabric filters, he or she is referred to the vast body of literature on the subject covering everything from the basics to complex mathematical analogs (fabric filters are fairly simple things compared to, say, an automobile, and one should be able to acquire a basic body of knowledge easily).

Before launching into a discussion of the five main objectives, it might be useful to point out that specifications do not have to be formal written documents. On the contrary, they may be no more than a telephone conversation with the local representative of a pollution control manufacturer, although, as pointed out elsewhere, there may be real advantages to setting specifications down in writing.

As a preliminary matter, specifications should always state the following:

(1) Company name
(2) Street address
(3) City, state, zip code
(4) Phone number
(5) Contact's name and title

Note well: Items 1 though 5 should be supplied for both the corporate office and the job site if they are different.

(6) Bid deadline
(7) Firm bid or budget estimate

6.2 SYSTEM PARAMETERS

No one knows more about the process or operation to be controlled than the buyer of the baghouse, and no one knows more about the baghouse than the seller. A fundamental purpose of writing specifications is to convey information about the process to the seller so that he or she can make intelligent choices with respect to the design of the pollution control system in general and the baghouse in particular.

Although an exhaustive listing of pertinent system parameters is not possible, the following should provide a good beginning:

(1) Basic process description
(2) Continuous or intermittent?
(3) Can process be shut down for maintenance?
(4) Gas or air volumetric flow rate:
   – Maximum

– Minimum
– Normal

(5) Gas or air temperature:
– Maximum
– Minimum
– Normal

(6) Particulate loading:
– Maximum
– Minimum
– Design

(7) Approximate chemical composition of particulate (if inert, simply so state)

(8) Approximate particle size distribution (especially if significant percentage by weight is less than 3 μm in diameter)

(9) Approximate chemical composition of gas stream (especially if water vapor or corrosive species are present)

(10) Is dust combustible, or is aerosol explosive?

(11) Fuel analysis (especially sulfur and chlorine) if process involves combustion

6.3 LIMITING THE SCOPE OF THE SELLER'S PERFORMANCE

Most large manufacturers of air pollution control equipment are competent to accept a contract for the most sophisticated turnkey system – or for the simplest dust collector.

Assuming that the buyer's management has already decided what the scope of the seller's performance will be, the specifications should clearly convey the substance of the decision to the seller.

Even if a turnkey system is not desired, it might still be advantageous to purchase system components other than the basic baghouse from the baghouse manufacturer. Fans, motors, drives, air compressors, switchgear, and control panels are commonly purchased from the baghouse manufacturer. The following are the advantages:

(1) Convenience; fewer sources to deal with

(2) Reduced lead time (assuming manufacturer inventories these components)

(3) Focuses responsibility; less opportunity for "finger pointing" if (when) something goes wrong

The big disadvantage is, of course, cost. These components will be more expensive if purchased from the baghouse manufacturer because of the markup and, assuming these items are inventoried, the carrying charges.

If a full-blown turnkey system is desired, a threshold decision must be made as to whom the buyer wants as the seller. The usual choice is between a baghouse manufacturer and a general contractor who purchases the equipment from the baghouse manufacturer and other suppliers and subcontracts the various erection services or will subcontract the entire erection portion of the main contract including the general contracting function.

Buyers usually feel most comfortable dealing with the baghouse manufacturer. This is so not because the baghouse is more costly than the erection services (it may well be less) but because in the final analysis it is the baghouse design and manufacturer that is dispositive of the success of the system and the warranties extended to the buyer by the seller the primary safeguard of this success. Admittedly, warranties may be made by the baghouse manufacturer to a general contractor who may then pass them along to the ultimate user, but buyers tend to feel most comfortable without the buffer between themselves and the ultimate maker of the warranties.

It is not uncommon for manufacturers to do a substantial amount of free engineering in order to bid on a project. Of course the engineering is not really free since the manufacturer's price will ultimately reflect this engineering, and the amount of engineering a manufacturer will do probably depends on the size of the job as well as the manufacturer's assessment of how likely it is that he or she will ultimately be awarded the contract. Many buyers are masters at extracting engineering from eager sellers, and it is not uncommon for specifications to be extremely sketchy and include instructions that make it necessary for bidders to visit the job site prior to bidding to develop for themselves the system information necessary for bidding. This approach also tends to provide a sound basis for the Uniform Commercial Code (UCC) that implies warranty of "fitness for a particular purpose." When this approach is taken, it is probably more accurate to call the buyer's document an *invitation to bid* rather than *specifications*.

Even though a buyer purports to require a turnkey contract, very often this is not literally true. Many buyers prefer to provide specific items themselves, usually because they are specially equipped to do so or because they feel they are not likely to be satisfied unless they do it themselves. The following are items which frequently fall into this category:

(1) Foundations, anchor bolts, and grouting
(2) Field painting
(3) Payment of local taxes
(4) External wiring
(5) Motors above a certain horsepower

(6) Control panels
(7) Insulation
(8) Lighting
(9) Electrical switchgear
(10) Collected dust storage and/or disposal system

6.4 INFORMING THE MANUFACTURER ABOUT WHAT THE BUYER WISHES TO BUY

6.4.1 SYSTEM CONFIGURATION

Regardless of whether the buyer wishes a turnkey system or baghouse equipment only, he or she may have very definite wishes as to how the components should physically relate to each other. This, in turn, will very often dictate important details of construction or energy requirements. Thus it becomes crucial to communicate the planned system configuration to the manufacturer. Several fundamental aspects of system configuration will be discussed.

Whether the baghouse is located on the inlet or discharge side of the fan is of elementary importance. In general, baghouses located on the discharge or pressure side of the fan are cheaper because they are usually subjected to less pressure differential; the absence of stacks and a considerable portion of ductwork further decreases first cost. The main disadvantages are considered to be increased maintenance on the fan due to handling dirty gas and difficulty in assessing baghouse collection efficiency due to the absence of stacks.

Other choices of component or location of component decisions that should be communicated to the manufacturer are the following:

(1) *Baghouse location*: If a particular location for the baghouse has been selected, a plant drawing indicating the location should be included with the specifications.

(2) *Controls location*: Will the controls be located in a simple panel mounted on, or close to, the baghouse, or will the controls be incorporated into the main control room serving the process? Very often a separate control room is built for the baghouse system if the system is complex.

(3) *Dust disposal*: The method of dust disposal will often dictate hopper design and the type of hopper discharge device to be used. The common types of dust disposal systems are screw conveyors, pneumatic conveyors, and wet sluice systems.

(4) *Gas cooling*: The method chosen for cooling hot gas to temperatures acceptable to filtration media (that is, less then 550°F for fiberglass and lower still for other fabrics) profoundly affects the volume, dust concentration, and humidity of the gas stream.
(5) *Utilities*: The location and capacity of existing utilities such as electrical power, water, and compressed air may play a large role in the selection of the type and location of the control device. Utility location and approximate capacity should always be indicated on the plant drawing.

Even in a simple dust collector, a buyer may often have specific preferences with respect to construction details. It should be recognized that they may be more expensive than they're worth if they cause the manufacturer to deviate substantially from normal fabrication methods. Construction details often specified for small dust collectors are the following:

(1) 16- , 12-, or 10-gage steel construction
(2) Factory-installed bags
(3) Specified bag material
(4) Interior prime and finish coat of paint
(5) Factory-installed screw conveyor
(6) Specified NEMA construction for electrical motors and switchgear
(7) Specified hopper dust discharge device
(8) Factory wiring
(9) Stainless steel bag clamps with a specified type of locking device
(10) Specified hopper clearance
(11) Hopper angle, 60° minimum
(12) Specified manufacturers for components supplied but not manufactured by the baghouse manufacturer

It is possible to have much more flexibility in specifying construction details when purchasing a large custom-made baghouse. It should be kept in mind, however, that specifying details probably has the effect of waiving the UCC-implied warranty of "fitness for a particular purpose."

Construction details often specified on larger systems are as follows:

(1) Structural design to withstand specified wind loads or snow loads or to be designed for a specified seismic zone rating
(2) Housing and hoppers to be fabricated of a specified thickness of metal and to be stiffened to withstand a specified pressure differential (for example, 3/16-in. mild steel plate housing and 1/4-in. mild steel plate hoppers; stiffened for 30 in. of water column pressure differential)

(3) Hopper equipped with clean-out pipes; strike plates and sized to accommodate a specified quantity of collected material
(4) Specified hopper clearance
(5) Specified insulation material, thickness, and lagging
(6) Specified paint
(7) Hopper heat via electrical resistance or inductance
(8) Specified bag material, size, spacing, and reach. Probably the most common bag arrangement on large projects is 12-in.-diameter by 30-ft-long bags on approximately 14-in. centers with a two- or three-bag reach from the maintenance walkway
(9) Specified damper construction. Air cylinder operated poppet valves are the most common compartment dampers with vertically acting guillotine valves the most common for system shutoff or bypass.
(10) Specified bag cleaning method (that is, shake, reverse air, pulse, or a combination of these)

It is very common to insist that manufacturers bid according to the specifications and submit alternate bids to incorporate design features that the manufacturer feels are superior to, or have advantages over, those specified.

When specifying construction features in great detail, it is important to avoid writing the specifications so tightly around a particular manufacturer's equipment that other sellers will be discouraged from bidding (or will merely submit a token bid). Of course, it might be that management has virtually decided that the construction details offered by manufacturer A are the only ones acceptable, in which case the advantage of writing specifications in the first place might be questioned. It is frequently said that even though a particular manufacturer will almost certainly be awarded the contract, management insists on "price checking" bids from competitors.

6.4.2 INTERCOMPARISON OF BIDS

It is patently obvious that the system to purchase is the one that will cost the least and is otherwise satisfactory. It is equally obvious that the system that will cost the least is not necessarily the system with the lowest purchase price.

The expenditure of several hundred thousand to several million dollars for an air pollution control system over its useful life warrants considerable effort in evaluation of the bids. This evaluation must take into account many cost variables other than first cost, and a well-structured set of specifications can make this evaluation much easier.

6.4.3 NEGOTIATING TERMS AND CONDITIONS

The buyer usually fires the first volley in the "battle of the forms" with his or her specifications. It is common for specifications to include a listing of warranties desired by the buyer, acceptable terms and conditions of payment, and the requirement that the successful bidder obtain a performance bond. Although it is extremely unlikely that the terms and conditions contained in the buyer's specifications will be acceptable to the seller as they stand, at least they provide a point of departure.

The common express warranties extended on baghouse air pollution control systems are described below along with typical language.

The common warranties that buyers attempt to obtain from the pollution control equipment manufacturers are the following:

(1) Emission warranty
(2) Pressure drop warranty
(3) Reliability warranty
(4) Warrantied start-up date
(5) Material and workmanship warranty (especially if implied warranties are disclaimed)
(6) Bag life warranty
(7) Performance bond

Any discussion of warranties must necessarily focus on remedies for breach of warranty. *Sole and exclusive* remedies are generally inserted into the final contract and stated as *liquidated damages* to circumvent the illegality of penalty clauses.

The raison d'etre for spending millions of dollars for a control system is simply to achieve the legal emission rate. Manufacturers universally make emission warranties, and the following wording is typical:

> Seller warrants that the system as quoted in this proposal will provide an emission in compliance with all applicable state, local, and federal codes for solid particulate emission in effect as of the date of this proposal. In addition, the outlet emission is warranted to be less than 0.005 gr/scf and the opacity of the emission to be less than 20% neglecting any condensed vapors.
>
> This warranty is based on the system being operated on the application and at the air volume as described in the project specifications. Also, this warranty is based on the baghouse being operated properly and in accordance with seller's instructions.
>
> This warranty will be in effect until a properly qualified independent testing company has performed an emission test which shows compliance, the governing air pollution agency has accepted this installation, or 6 months

after the date of initial operation, whichever occurs first. Seller has the right to be notified of an emission testing schedule 1 week prior to actual testing, to be present during emission testing, and to review emission test data. Should the system fail to perform as warrantied, buyer must promptly notify seller in writing of such failure, and seller shall have a reasonable time within which to make any repairs or modifications it desires to bring the system into compliance.

It is the manufacturer's intention that the system will operate properly and in accordance with all express warranties. However, if the manufacturer is unable within a reasonable time to make the system perform as warrantied, or is unwilling to do so, then buyer's sole and exclusive remedy for breach of the above warranty shall be the refund to buyer by seller of an amount of money not to exceed 20% of the original price of the equipment (not including field erection, insulation, wiring, and piping) less the moneys already spent by the manufacturer to attempt to repair the system.

The fan power cost is directly proportional to the control system pressure drop.

Manufacturers make pressure drop warranties, and the following wording is typical:

Seller warrants that the system as quoted in this proposal shall cause a static pressure drop of no greater than 5.0 in. water column, as measured in the process gas flow at the outlet manifold flange with a reference point in the process air flow at the inlet manifold flange, when the equipment is operated at design conditions.

Also, this warranty is based on the system being operated properly and in accordance with seller's instructions.

This warranty is in effect for a period of 6 months from date of initial operation or until March 1, 19XX, whichever occurs first. Seller has the right to be notified of testing schedule 1 week prior to actual testing, to be present at testing, and to review test data.

Should the system fail to perform as warrantied, buyer must promptly notify seller in writing of such failure, and seller shall have a reasonable time to bring the system into compliance.

It is the manufacturer's intention that the system will operate properly and in accordance with all above warranties. However, if the manufacturer is unable within a reasonable time to make the system perform as warrantied, or is unwilling to do so, then buyer's sole and exclusive remedy for breach of the above warranty shall be the refund to buyer by seller of an amount of money not to exceed 20% of the original price of the equipment (not including field erection, insulation, wiring, and piping) less the moneys already spent by the manufacturer to attempt to repair the system.

An unplanned outage is an extremely costly occurrence. Buyers attempt to insure themselves against the monetary loss occasioned by such an event, if attributable to a malfunction of the pollution system, by the device of a reliability warranty.

Most manufacturers are reluctant to make this warranty, and at least one prominent manufacturer of both electrostatic precipitators (ESPs) and baghouses stated that although they would make such a warranty (98% availability for an ESP and 99% for a baghouse), they would exercise great care in specifying the remedy and would never offer the warranty unless demanded by the buyer. Another prominent manufacturer of both baghouses and precipitators stated that they would be "extremely" reluctant to make a reliability guarantee. Reasons given for this reluctance are the following: (1) lack of control over, and difficulty of determining, specific cause of outage and (2) exposure to extremely large monetary loss.

The successful bidder on a recent large baghouse project made a reliability warranty with the sole and exclusive remedy for breach being the payment of $3500 as liquidated damages for "extra fuel." This remedy is applicable for "any outage caused by the system."

Remedies for breach of warrantied start-up dates are very difficult to distinguish from penalties. The successful bidder on the project mentioned above made a start-up date warranty with $500,000 being the limitation on "actual" damages for breach.

As discussed in more detail elsewhere, the Uniform Commercial Code (UCC), which is law in every state, provides for certain "implied" warranties; these warranties are made by a manufacturer to a buyer, by operation of law, unless they are specifically disclaimed, in writing, by the manufacturer. In other words, the manufacturer makes these warranties, whether he or she intends them or not, unless it is specifically stated in the written contract that they are not made.

The most important implied warranties under the UCC are the following: (1) merchantability, that is, the goods sold are fit for the ordinary purpose for which such goods are sold, and (2) fitness for a particular purpose, that is, in addition to being merchantable, the goods are fit for any purpose for which the seller has actual knowledge and, in addition, the seller knows that the buyer is relying on the seller's skill in this area.

Unless modified by contractual language, material and workmanship fall within the ambit of merchantability. Virtually all manufacturers attempt to disclaim the implied warranties and substitute an express warranty with a limited remedy. Here is a typical example:

Seller warrants that the material and services supplied are free from defects in material and workmanship for a period of 1 year from date of operation or until August 1, 19XX, whichever occurs first. This warranty shall be expressly in lieu of any other warranty, expressed or implied, including, but not limited to, any implied warranty of merchantability or fitness for a particular purpose.

Bag life warranties used to be universally disclaimed by vendors. Now the opposite is true, and virtually all manufacturers offer bag life warranties, at least on large projects. Six of the seven bidders on the recent project mentioned above offered a prorated bag life warranty. A typical bag life warranty reads as follows:

> Seller warrants that the baghouse system bags as described in this proposal will have a life of 24 months, excluding a normal replacement rate of 15%. This warranty is based on the baghouse being operated on the application and at the air volume and temperature as stated in the proposal. Also this warranty is based on the baghouse being operated properly and in accordance with seller's instructions. This warranty will be in effect for 24 months from the date of initial operation or until August 1, 19XX, whichever occurs first. In the event that more than 15% of the bags fail during the warranty period, the buyer will be credited or refunded (at seller's option) an amount equal to the original selling price multiplied by the percentage of warrantied time use not received by buyer. Seller shall have the right, during the warranty period, to make any modifications it desires, without hindering cleaning efficiency, to improve bag life.

Sellers of large air pollution control systems are routinely required to be bonded for performance by the buyer. This is essentially a third-party beneficiary contract and, in effect, operates to give the buyer a solvent party to sue (the indemnitor) in the event the seller is unable to perform to completion after the contract has been awarded.

6.5 FABRIC FILTER WITH DRY SCRUBBER

Some of the more advanced systems utilize a dry scrubber followed by a dry fabric filter arrangement. This section discusses such a system, giving performance specifications for lime injection units. Note that the example uses this as the air pollution control system for a medical waste incinerator.

6.5.1 GENERAL

(1) The CONTRACTOR shall provide and install a dry chemical scrubber air pollution control system specifically designed for control of particulate and acid gas emissions from an infectious waste incinerator.

(2) The CONTRACTOR shall submit proposals for systems designed for two different size requirements.

System A – Designed for operation with the current incinerator.

System B – Designed for operation with the current incinerator as well as a future incinerator to be located at a site remote to the hospital. This future system would serve not only the hospital and medical center, but also a variety of campus facilities.

Complete operating requirements are detailed for each system in further paragraphs. Given the gas flow rates specified and the available equipment sizes, if it is possible to simply add components to System A to achieve the capacity required for System B, then System B could be proposed as an add-on cost to a base price for System A. If this is not feasible, then each system should be proposed as an independent bid with a separate price.

(3) If System B is selected, it is anticipated that the equipment would be installed at the current incinerator location and then moved to a remote site for operation with the future incinerator. The CONTRACTOR shall submit, along with his bid for System B, an estimated cost for dismantling the equipment and reinstalling it at another location.

(4) A process flow diagram representing the current incineration system is included as Figure 6.1.

6.5.2 HYDRATED LIME STORAGE

(1) Lime may be introduced to the system in either a dry or slurry form.

(2) A single storage silo shall be provided to serve the air pollution control system and sized for 7 days storage capacity, single compartment. Roof shall be designed to support all roof mounted equipment plus a superimposed live load of 50 lbs per square foot. Silo shall have 60 degrees cone bottom. Silo shall be furnished with connections in the straight section for low level bin signal connection and high level bin signal. The silo and accessories shall be designed for an outdoor installation.

(3) Roof shall be furnished with a 20 in. diameter combination manhole inspection port including dust-tight closure with locking hasp; pressure/vacuum relief valve; dust filter adapter; and air filled turbulence discharge box.

(4) Storage section shall be fabricated of minimum 3/16 in. plate and equipment section shall be fabricated of minimum 1/4 in. plate. All steel plates shall be ASTM A36. Bottom flange shall include three leveling screws and anchor bolt saddles.

(5) Equipment section underneath shall be furnished with two 36 × 84 in. access doors with hardware; and two vents with bird screens.

(6) Complete storage and equipment section shall be of welded construction, shipped in one piece, and furnished with lifting lugs.

(7) The storage silo shall be furnished with an external ladder complete with cage, platforms at bin level indicator and at intervals not exceeding 30 ft, and steel support members. Platforms shall be provided with

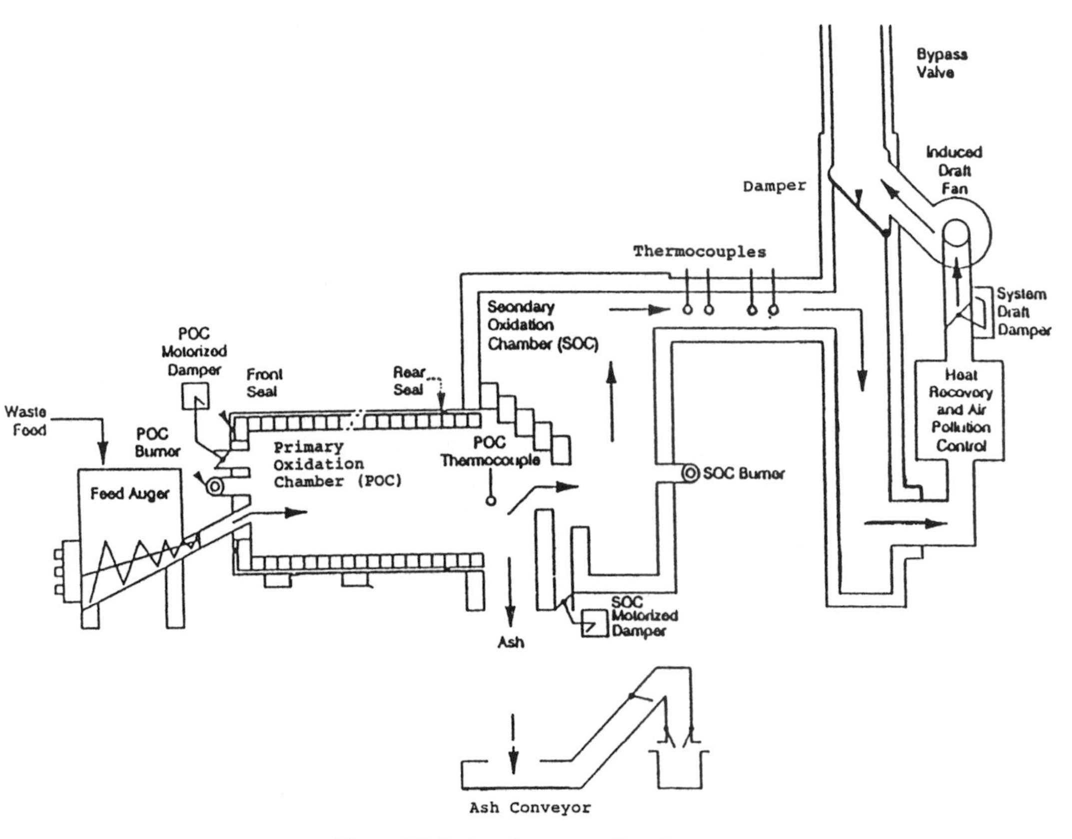

Figure 6.1 Incinerator process flow diagram.

handrails and toe plates. Handrails and toe plates shall be provided around the perimeter of the silo roof.

(8) The silo discharge will have a motorized bin activator (5 ft min. dia.) to prevent bridging, jamming, and segregation and to ensure positive flow of material on a first-in, first-out basis. Design of the bin activator discharger shall prevent transmission of vibrations to the storage silo. Bin activator shall be of the following materials of construction:
 - –contact materials: carbon steel
 - –external support brackets: carbon steel
 - –flexible screws: suitable reinforced natural or man-made elastomer
 - –motor: 1.5 HP, TEFC, 208 volt, 60 hertz, 3 phase
 - –outlet size: shall be determined by the manufacturer, stated in the bid and approved by the ENGINEER

(9) Two level indicators shall be furnished, one for high level and one for low level. Indicators shall be motor operated type with stainless steel paddle or vibrating probe. High level indicator shall be furnished with 3 in. shaft extension on silo roof. Low level indicator shall be furnished with paddle guard adjacent ladder in straight walls of silo. Indicators shall be accessible from the ladder and pre-mounted prior to shipment.

(10) Storage silo fill pipe shall be 4 in. dia., schedule 40 seamless steel furnished complete with 4 ft radius elbows, flanges, truck connection, dust cap with chain, and limit switch on end of pipe to provide automatic operation of the silo vent filter during and after truck unloading cycle. Limit switch assembly shall be adjustable and secured to a heavy-duty mounting bracket. Limit switch shall be complete with one set of N.O. and one set of N.C. contact. Spare contacts shall be wired into the unloading control panel to numbered terminal strips for remote monitoring of filling operation. Fill pipe shall extend through outside plant wall and shall terminate approximately 5 ft 0 in. above grade at the truck unloading/silo fill station and shall be shipped unassembled for field installation.

(11) Silo shall be furnished with a roof mounted dust filter with retained dust to discharge directly into storage bin. Filter shall have a suitable cloth area to handle 1200 scfm. Filter assembly shall be in a steel housing and weather proof with gasketed access doors.

(12) The storage silo shall be furnished with a truck fill panel mounted near the truck unloading/silo fill station. Truck fill panel shall be in a NEMA 3R enclosure. Panel shall include:
 - –selector switch, key operated, for power ON/OFF
 - –push button for manual dust filter pulse operation

–push button for alarm silence
–indicating lights for: Power "ON," dust filter pulse "ON," high bin signal, low bin signal
–alarm horn

(13) The lime storage silo shall be located as shown in the Facility Plot Plan.

6.5.3 LIME FEEDER

Lime feeder shall be volumetric dry solids type screw feeder with adjustable capacity for the range required and suitable for handling hydrated lime with a fineness varying from 75% passing a #200 sieve to 99% passing a #325 sieve. Flow into the feeder shall be through a 3 cu ft feed hopper equipped with a vibrator. Feeder shall be complete with a full wave SCR controlled with 4 to 20 maDC proportional pacing DC variable speed drive having a 30:1 output range. All speed adjustments shall be stepless and can be made with the feeder operating. Controller shall be in a water-tight, dust-tight NEMA 4 enclosure mounted adjacent to the main control panel, and shall include a digital speed selector, ON-OFF switch, line fuse and armature fuse. The drive motor shall be TEFC, 1800 RPM direct coupled to 30:1 right angle gear reducer.

6.5.4 LIME TRANSFER

(1) A motor operated rotary vane feeder shall be provided to form an airlock. It will be designed to be leak-proof at up to 150% of the pressure differential expected.
(2) The CONTRACTOR shall provide a pneumatic dry sorbent transfer system including a positive displacement blower, 3 in. sch 80 conveying pipes, long radius elbows, fittings, compilings and pipe supports. All structural support required for transfer system from storage silo to scrubber location shall be provided by the CONTRACTOR.

6.5.5 HEAT EXCHANGER

A heat exchanger shall be provided to cool flue gases from the waste heat recovery steam generator to enhance the acid gas reaction. Preferred design type is reverse shell-and-tube; flue gas on outside of cooling tubes, cooling air blown through insides of tubes.

(1) Cooling air shall be automatically controlled to maintain outlet flue gas temperature of a minimum of 30°F above the flue gas saturation temperature.

(2) Maximum allowable gas outlet temperature variation is plus or minus 10°F from set point.

(3) Maximum allowable pressure drop across heat exchanger at maximum charging rate operating conditions is 3 in. wc.

(4) Heat exchanger shall be equipped with soot blowers and/or sonic horns for automatic cleaning of cooling tubes at adjustable time intervals.

(5) Horizontally mounted, replaceable cooling tubes with Vision seals are to be provided.

(6) Gas inlet is to be at the top and the gas outlet is to be at the bottom of the unit.

(7) In the bid give special consideration to the difficult nature of this application and state method of protecting against condensation and associated operational problems. State tube material and shell construction and lining.

6.5.6 REAGENT INJECTION AND REACTION CHAMBER

Hydrated lime shall be injected through a nozzle designed to evenly disperse the hydrated lime across the flue gas stream. A variable throat Venturi will be provided upstream of the nozzle to minimize fall-out during low load conditions.

(1) Lime injection will be between the heat exchanger and a reaction chamber.

(2) CONTRACTOR is to supply a reaction chamber if required to get suitable reaction time for efficient reagent usage. Construction is to be minimum 3/16, plate and complete with hopper for any fall-out. Provide in bid an analysis demonstrating that design provides sufficient reaction time and state reagent usage rate used as a design basis.

6.5.7 FABRIC FILTER

The baghouse shall be an outside bag collection type with pulse jet cleaning and shall consist of a complete, fully assembled module, and shall be shipped with hopper attached.

(1) Housing and Tube Sheet: The housing and tube sheets shall be a minimum of 3/16 in. thick and of all welded construction. Tube sheets shall be seal welded to the module walls and shall include stiffeners to limit the deflection of the tube sheet under loading due to the bags, cages, Venturis, dust load and foot traffic to a maximum of 0.015 in. per foot. The arrangement shall be such as to permit individual

removal of bags and cages from the top of the fabric filter. Interior surfaces that are exposed to gases that have passed through the fabric filters shall be sandblasted and epoxy coated.

(2) A dust hopper shall be included and shall be sized for a minimum of four hours storage time based on 30 pounds per cubic foot ash density. Structural design and loadings shall be based on a density of 100 pounds per cubic foot based on the hopper full of ash. The hopper shall have a minimum valley angle of 55 degrees. It shall be a pyramidal type and shall be constructed of 3/16 in. minimum carbon steel plate.
- –The CONTRACTOR shall furnish one six-inch-square strike plate on the hopper. Vibrators shall be provided as well as one 4 in. dia. rod-out port.
- –The CONTRACTOR shall provide one vibrating type level detector for indication of high hopper dust level.

(3) Bags and Cages
- –Bags and Cages will be cylindrical, nominally 6 in. dia. and mounted vertically, and not exceed 12 ft in length.
- –Bag cages shall be fabricated in one piece. Split cages are not acceptable. Each cage shall be constructed of wire mesh or rigid wires to provide suitable bag support. The cage will have a solid bottom pan and an integral Venturi or nozzle at the top. All cage materials shall be carbon steel. Finished cage shall be of rigid construction and shall be cylindrically smooth and straight throughout its full length. Special care shall be taken to assure there are no rough spots on cages to cause bag abrasion.
- –Cage attachment to tube sheet shall be rigid to minimize movement of bags during operation.
- –Bag material shall be polyimide (trade name P-84) and should have a maximum air to cloth ratio of 5:1.
- –Bag and cage design shall be such that bag length will be fitted to cage length and no fabric folds will exist on installed bags. CONTRACTOR shall describe design used to accomplish this. Any exposed bands or clamps used shall be stainless steel. A double thickness wear strip shall be provided at the bottom of the bags to resist abrasion from bags hitting each other.
- –Bag life shall be guaranteed for a minimum of two years.

(4) Bag Cleaning Equipment
- –The baghouse shall have a factory installed compressed air manifold. The manifold shall consist of a steel pipe or square tube, mounted external to and running the full length of the unit. The manifold shall include a drain port.

–Between the air manifold and blow pipes for each row of bags, shall be a double diaphragm type air pulse valve. Each air pulse valve shall be controlled by a direct connected 110 volt, single phase, 60 hertz, solenoid pilot valve. Remote solenoid valves connected by tubing will not be permitted. Solenoid valves shall be pre-wired to a common junction box.

–Blow pipes shall be provided above rows of bags. A maximum of 16 bags may be pulsed from each flow pipe and air pulse valve. Blow pipe shall be fabricated of Schedule 40 steel pipe with orifice holes located accurately over Venturis. Blow pipe system shall be designed and rigidly supported at each end to maintain accurate lateral and vertical orifice hole/Venturi alignment and spacing. Valves, piping, wiring, etc., shall be protected from weather and traffic and are to be easily accessible for maintenance.

–The system shall include an air compressor sized to provide sufficient pressure and flow to bag cleaning equipment proposed.

(5) Tube Sheet Bag and Cage Access

–CONTRACTOR shall provide pre-insulated lift-off or hinged top door(s) for bag access. Door shall incorporate a 3 in. layer of thermal insulation between a checkerplate top suitable for foot traffic and 3/16 in. steel plate on the underside. Doors shall be secured by hold down clamps and shall have high temperature seals to minimize flue gas leakage or air infiltration and provide a weatherproof seal with the module.

(6) Fire Protection

–The CONTRACTOR shall provide heat suppression and flame detection equipment, for fire protection of the proposed system.

6.5.8 DUCT AND DAMPERS

(1) Provide ductwork interconnecting each piece of equipment supplied and connecting from the existing system to the proposed system, and to the existing stack.

(2) All duct is to be a minimum of 3/16 in. stainless steel plate and sized for a nominal 3600 fpm velocity at design conditions.

(3) The CONTRACTOR shall include an analysis or model study of duct and equipment to assure good dispersion of reagent and minimum reagent/ash fall-out.

(4) Damper Valves:

–The module will have both an inlet and an outlet isolation valve. Both shall be pneumatically operated butterfly valves with provisions to manually lock them in the closed position. Each damper shall be shop assembled and tested and shall be equipped with flanges for connection to the inlet and outlet. They shall be provided with limit switches to indicate both full open and full closed positions. Limit switches shall have NEMA 3R enclosures and DPDT contacts.

–A double bladed poppet valve is to be provided for system by-pass from the inlet to the outlet. There will be two (2) sealing plates to assure 100% isolation. This valve will have properly stiffened stainless steel valve plate and stainless steel shaft. The valve is to operate in a vertical position. Damper shall be operated by a pneumatic actuator with provisions for locking the valve in position. Provisions shall be made for field alignment of the operator. Damper operating speed shall be completely adjustable during both opening and closing by means of a speed adjusting mechanism. Actuator shall be provided with flexible air hose(s), air filter and lubricator. Regulators shall be provided if required. The damper shall be provided with limit switches to indicate both full open and full closed positions. Limit switches shall have NEMA 3R enclosures and DPDT contacts.

(5) A by-pass duct connection between the inlet and outlet will be provided.

6.5.9 SUPPORT STEEL AND PLATFORMS

(1) The CONTRACTOR shall provide supports for all provided equipment. The support steel should have a level tolerance of plus or minus 1/8 in. The structural support steel shall be plumb, square, and aligned in accordance with AISC Chapter 5, Section 7.11.

–The fabric filter will have supports to provide for a clearance of 5′–6″ below the hopper flange.

–If the heat exchanger or reaction chamber has a hopper for ash fall-out, provide supports for a clearance of 5′–6′ below the hopper flange.

(2) Access facilities shall be provided to permit access to items of equipment which require maintenance, servicing, or inspection. Access doors shall be provided for internal access to the baghouse and components described below.

–Service platforms shall be provided at any level for access to all module dampers, operators, bag cleaning systems, and access doors. Valves, piping, wiring, and the like, shall be easily accessible for maintenance from an adequate walkway.

– Service platforms shall also be supplied for any equipment which may require routine maintenance including heat exchanger fan, lime injection equipment, lime silo accessories, etc., unless equipment is sufficiently near the ground and infrequently serviced to make it practical to use temporary ladders. Provide in bid location of all service platform included with the system.
– Access to all service walkways from grade shall consist of caged ladder(s).
– Walkways, ladders, stairways, grating and handrails are to be designed per Federal and State OSHA requirements.

6.5.10 AUXILIARY HEATING AND INSULATION

The CONTRACTOR shall provide an auxiliary heater and insulation designed to maintain a minimum system operating temperature of 20°F above the saturation temperature of the gas stream. The CONTRACTOR shall include in the proposal a design analysis to show that the equipment proposed will meet this performance criteria.

6.5.11 CONTROL SYSTEM AND INSTRUMENTATION

(1) A complete integrated system of instruments and controls shall be provided to measure, monitor, operate, and control the system. The air pollution control system shall be operated automatically from a main control panel located near the baghouse. This panel shall be complete and shall house all instruments, controllers, indicators, relays, power supplies, fuses, indicating lights, push buttons, switches, transformers, contactors, programmable controller, alarm contact, and other devices as required to provide a complete and operable system consistent with the intended operation as described herein.
(2) The CONTRACTOR shall provide all required local instruments, transmitters, sensing probes.
(3) Instrumentation power shall be 120 VAC, 1 phase, 60 Hz.
(4) Instrumentation control signal shall be 4–20 MA DC.
(5) Instrumentation enclosures shall be NEMA 3R.
(6) Switches and contacts shall be DPDT with dry contact to operate at 24 VDC or 120 VAC.
(7) All electrical equipment components such as, but not limited to, switches, contacts, relays, enclosures, powered instruments, etc., shall be listed by Underwriters Laboratories (UL). If required to do so, the CONTRACTOR shall provide the Company proof of UL

listing of a component(s) in the form of a photocopy of the UL guide card for the component(s) in question.

(8) Cabinet Structure

–Panels shall be NEMA 3R, free standing or wall mounted design for front access and fully enclosed construction. Louvers, exhaust fans, and heaters shall be provided, if required, to maintain electronic instrumentation within CONTRACTOR'S specified operating temperature range.

–All internal panel electrical components shall be open type (not enclosed). All panel wiring shall be open type, neatly bundled and supported, or semi-open, run in plastic wiring troughs, and shall be provided with unique wire number markers (plastic sleeve type). All wiring requiring external connections to the panel shall be run to terminals properly marked as per the approved shop drawings for external connections. Hand lettered or adhesive type terminal identification is not acceptable. Not more than two field connected wires shall be terminated on any one point.

(9) The system shall be operated, monitored, and controlled by a single main microprocessor-based programmable logic controller (PLC), located in the control panel. Push buttons, selector switches, and field device signals from limit switches, timers and differential pressure transmitter shall be handled as inputs to the PLC logic, which shall control all necessary interlocks, operating logic, signal interpretation, and field devices. The control system shall monitor and operate the heat exchanger, lime injection and baghouse cleaning.

–Heat Exchanger Controls: The control panel will modulate flow of cooling fluid to the heat exchanger to maintain the set point outlet temperature.

–Lime Injection: There will be a manual adjustment for rate of lime injection. In addition, an alternate automatic mode is to be available which will accept a customer-supplied 4–20 m.a. signal for adjustment of flow.

–Baghouse system cleaning shall be initiated by pressure differential across the system. There shall be an adjustable timer for secondary initiation of cleaning. Once cleaning has started, it will continue until the pressure drop across the bags reaches an acceptable predetermined value.

The following features and functions shall be incorporated into the system.

(a) Cleaning times (duration and interval) shall be adjustable.
(b) Differential indicator/switch for baghouse differential pres-

sure from inlet to outlet shall be provided. Inputs from differential switch shall be used by PC for control of module cleaning operations as required.

(c) Inlet, outlet, and by-pass damper position switches shall be used in the control logic for damper positions failure and to indicate when the system is off-line for servicing.

(d) Chart Recorder: One (1) two pen, 12 in. dia. circular linear chart, internally illuminated, hinged front, disposable filter type pans. Include 400 charts of 24 hour duration and two (2) spare sets of pens. Recorder shall record inlet gas temperature and fabric filter differential pressure.

(e) Baghouse inlet RTD with well mounted in inlet manifold complete with temperature transmitter if required. Temperature high switch shall also be provided. The RTD is to be 100 OHM platinum type (at zero degree C) with 304 stainless steel sheath, standard 3-wire lead, and weatherproof connector lead.

(f) Compressed air low pressure switch with separate dry contacts for remote sensing.

– Graph Panel Indicating Lights:
The control panel face shall have a graph panel of the system with lights at appropriate positions to indicate:

(a) Cooling fan on
(b) Lime system operations: Bin actuator on; Feeder on; Pneumatic blower on; Rotary valve on; High, low bin level
(c) High baghouse differential pressure
(d) Cleaning system on
(e) Cleaning system control power on
(f) Low compressed air pressure
(g) By-pass damper open/closed
(h) Inlet damper open/closed
(i) Outlet damper open/closed
(j) Other indicating lights as recommended by the Seller to properly monitor the baghouse system
(k) Baghouse differential pressure indicator
(l) Audible alarm to annunciate high differential pressure or high inlet temperature
(m) ID fan on

– Control panel face shall include all necessary selector switches and/or pushbuttons to initiate control and operating functions.

– Alarm conditions including, but not limited to, high inlet temperatures, low inlet temperatures, lime supply failure, high baghouse differential pressure, damper position failure, shall be configured to provide a single dry alarm contact for remote indication.

–Programmable Logic Controller (PLC):
The programmable logic controller shall include three basic components. These components are the processor, power supply, and input/output (I/O) section.

(a) The processor shall be a complete solid-state device designed with sufficient capacity for all required inputs and outputs. The main function of the processor will be to continuously monitor the status of all inputs and direct the status of all outputs. The processor will be designed for a hostile industrial environment and a maximum ambient operating temperature of 100°F. It will operate on 120 volts, ±15%. The memory shall be equipped with five year life lithium batteries to provide DC power to maintain memory whenever there is an extended power failure or power is turned off. The processor shall be provided with the following controls and indicators: Run light, power on light, battery on light, and memory protect key.

(b) The power supply shall be mounted inside the front cover of the processor and shall not require any adjustments or maintenance. The power supply shall have sufficient capacity to operate the processor and the required number of inputs and outputs.

(c) The input/output section shall consist of the various required types and numbers of I/O modules that will be rack mounted by using heavy duty metal housings designed to contain the I/O modules. Each I/O rack will be directly connected to the controller. The I/O modules will be solidly constructed units that are easily removed or plugged into their housing. Once inserted, electrical contact is automatically made through plated spring connectors. There will be no requirements to shut down the system to replace I/O modules.

(d) The PLC shall be factory programmed by the CONTRACTOR to accomplish the functions required for proper operation of the system.

6.5.12 INDUCED DRAFT FAN

(1) An induced draft fan shall be provided to maintain proper draft throughout the system.

(2) The fan should be designed to compensate for variations in outlet pressure from the APC system. The CONTRACTOR shall provide,

along with his bid, documentation of design features included to accomplish this.

(3) The induced draft fan shall be constructed such that all parts that contact the gas stream are constructed of Hastelloy C276 or equal.

(4) The fan shall be constructed with fan wheel hub extended through the fan backplate and a Teflon seal installed on the hub.

(5) Fan shall be statically and dynamically balanced.

(6) Fan shall be radial blade centrifugal. Backward inclined, and radial tip designs are not acceptable.

(7) A drain connection shall be provided consisting of a threaded pipe coupling welded to the lowest point on the scroll to allow condensate or other liquids to drain.

(8) An access door securely held by quick release latches shall be supplied for easy access to the wheel and housing interior for inspection or cleaning. The door shall be shaped to conform with the scroll curvature and gasketed to minimize gas leakage.

(9) The fan shall bear the AMCA seal. The rating of the fan shall be based on tests conducted in accordance with AMCA standards 210 and comply with the requirements of the AMCA certified ratings program.

(10) The CONTRACTOR shall submit, along with his bid, a complete set of draft calculations to demonstrate correct sizing of the I.D. fan. All assumptions and results should be clearly stated.

(11) A performance curve for the current induced draft fan shall be included.

6.5.13 STACK

(1) The system shall be designed to use the existing stack.

TABLE 6.1. Operating Data System A.

Exhaust gas flow rate	=3252 scfm
	=7979 acfm
Exhaust gas temperature	=699°F
Exhaust gas pressure	=29.64 in. Hg
Exhaust gas moisture	=9.7%
Carbon dioxide (CO_2)	=8.0%
Oxygen (O_2)	=12.9%
Carbon monoxide (CO)	=23 ppm
Nitrogen (N_2)	=79.1%
Particulate	=0.6527 gr/dscf @ 9% O_2
Hydrogen chloride	=1700 ppm

(2) The CONTRACTOR shall submit, along with his bid, an additional price per five feet section for extending the existing stack 20 feet.

6.5.14 PERFORMANCE REQUIREMENTS

(1) The system shall be designed such that the characteristics of the exhaust gas exiting the stack do not exceed the following criteria
- maximum temperature – 300°F
- maximum hydrogen chloride concentration – 50 ppm
- maximum particulate concentration – 0.015 gr/dscf @ 7% O_2

(2) The CONTRACTOR shall submit, along with his bid, the estimated control efficiency of the proposed system for the following compounds
- arsenic
- cadmium
- chromium III
- lead
- manganese
- nickel
- zinc
- benzene
- formaldehyde
- polynuclear aromatic hydrocarbons

(3) System A shall be designed based on the operating data provided in Table 6.1.

CHAPTER 7

Dry Scrubbers for the Control of Air Toxics

RICHARD P. BUNDY, P.E.[1]

7.1 INTRODUCTION

DRY scrubbers are proving to be one of the most effective methods to meet the new limitations on air toxics emissions required by the new Clean Air Act. Although there are several configurations of dry scrubbers, they all combine an acid gas scrubbing technique that produces a dry product with a particulate removal device for removal of all of the pollutants in a dry state. Although the particulate removal device can be either an electrostatic precipitator (ESP) or a fabric filter (baghouse), usually baghouses have been selected. The high efficiency that a baghouse provides, especially on sub-micron fumes, is the reason that many regulatory agencies have come to regard this technology as offering the best pollution control efficiency.

The earliest references to "dry scrubbing" occurred in the mid to late 1970s. At this time, the reference was generally for a system that combined a spray dryer with a fabric filter and was used on coal fired boiler applications for the simultaneous removal of SO_2 and particulate (fly ash). There were several systems installed at that time, but the establishment of new source performance standards (NSPS) for coal fired boilers mandated high enough SO_2 removal efficiencies that dry scrubbers of that era were not able to assure compliance. The expense of compliance with NSPS caused a significant drop-off in the number of coal fired boilers that were installed. Further, the marginal performance of the first dry scrubber systems and the popularity of the new fluidized bed combustion technology meant a virtual end of the market for dry scrubber systems.

[1]President, Bundy Environmental Technology, Inc., Reynoldsburg, Ohio.

In the late 1980s interest in these systems was revived because of the need to meet both acid gas removal and particulate standards for incinerator (municipal, industrial and regulated) applications. The new Clean Air Act requirements for control of specific toxic substances added a new dimension to the application.

7.2 TYPES OF DRY SCRUBBERS

Originally, "dry scrubbing" referred to systems that combined a spray dryer with a fabric filter. As time went on, a number of variations on that theme were developed and, although they are known by different names, they all are considered to be dry scrubbers. Following are some of the variations, alternative names that they are called, and a brief description of the operation of each.

(1) *Spray Dryer or Semi-Dry System*: A reagent is delivered in a slurry form to a spray dryer vessel where it is atomized into fine droplets. These droplets evaporate in the spray dryer and the reagent reacts with the acid gas in the wet form at the boundary of the droplet and continues to react in the dry state after it exits the spray dryer. The acids are converted to dry salts that are filtered, along with the ash and toxic materials, in a baghouse.

(2) *Dry Injection or Dry-Dry System*: The reagent is injected in a dry state before the baghouse and reacts with the acids as it is airborne and as part of the filter cake on the bags. All the solids are removed, as described above in the fabric filter.

(3) *Dry-Wet System*: In some cases, a wet scrubber is used after the baghouse to either improve the acid gas scrubbing, perform all of the acid scrubbing, and/or improve the condensation and removal of gaseous toxic materials. This wet scrubber can be added to either of the above dry scrubber arrangements, or can be used without any previous acid gas scrubbing (in this last case, "dry scrubbing" would be a misnomer).

7.3 DESIGN AND OPERATION

Each of the dry scrubbing techniques has three basic elements to the operation: temperature reduction and control, reagent injection, and particulate removal. In addition there are several subsystems that are not part of the basic environmental performance but are necessary for total system function. A summary of design alternatives is given in Table 7.1 and the rationale for selecting several are discussed below.

TABLE 7.1. Dry Scrubber Component Design Alternatives.

Gas Cooling System	Reagent Material	Reagent Injection	Particulate Removal
Heat recovery boiler	Activated carbon	Dry injection	Fabric filter
Evaporative cooler	Calcium based (lime)	Spray dryer	Electrostatic precipitator
Dilution air	Sodium based		
Heat exchanger			
Combinations			

7.3.1 TEMPERATURE REDUCTION/CONTROL

The flue gas from a combustion source typically exits at about 1800 to 2000°F, and has to be cooled for several reasons:

(1) To reduce the temperature from the combustion source to a level that the downstream equipment can tolerate. Typically a baghouse and other system components are limited to about 450–500°F.
(2) Acids are generally scrubbed more efficiently as the flue gas temperature approaches the wet bulb temperature.
(3) Some of the toxics exit as condensable gases. As the flue gas cools, these compounds condense and can then be removed by the fabric filter.

The type of gas cooling system that is selected can have an effect on the toxics emissions. Dioxins and furans are formed at mid-range temperatures and the faster that the flue gas passes through that range, the lower the formation of dioxins and furans. Also, the presence of iron oxide aids in the formation of dioxins, so exposure to clean, rusted boiler tubes will create higher levels of dioxins.

Different flue gas cooling techniques that can be used are noted in Table 7.2 and are discussed here.

(1) *Boiler*: A heat recovery boiler recovers heat by converting water to steam which can be used for plant requirements and/or sold to a nearby user. Corrosion is the greatest danger to a boiler system, and to avoid problems, the exit temperatures are not always brought as low as necessary; and a secondary gas cooling device is required.
(2) *Evaporative Cooler*: An evaporative or quench cooler reduces the temperature by the direct evaporation of water in the gas stream. This type system offers advantages in that the gas is cooled very quickly, which avoids the formation of dioxins. Evaporative cooling is an integral part of the spray dryer acid gas scrubbing system, and often

TABLE 7.2. **Alternative Gas Cooling Systems.**

Type	Advantages	Disadvantages
Boiler	Heat recovery Reliable No water carry-over risk Minimizes gas volume	High cost Disposal of steam req'd if not consumed May cause formation of dioxins
Evaporative cooler	Fast cooling inhibits toxic formation Reasonable cost	Risk of water carry-over Possible steam plume Adds gas volume (water) Consumes water
Dilution air	Low cost Small space required Simple	Adds considerable gas volume Major energy increase Dilution causes large correction factors Poor draft control
Heat exchanger	Reduces gas volume Easy draft control Small space required	Can only be used as a secondary cooling device Corrosion potential

evaporative cooling and spray drying are integrated into a single system.

Because the emissions will eventually enter a baghouse, it is imperative that any evaporative cooling device provide a totally dry discharge (complete evaporation of the water) to avoid blinding the baghouse.

(3) *Dilution Air*: On some systems, particularly small ones, it is feasible and cost-effective, to reduce the gas temperature by adding ambient air. This naturally increases the gas volume that must be treated considerably, and is not practical for larger systems.

(4) *Heat Exchanger*: A boiler as discussed above, is basically a heat exchanger; however, there are other types that are also used. Final gas cooling to a relatively low temperature can be done with a gas-to-water or gas-to-air heat exchanger. On smaller systems, the heat would not normally be recovered for any use. However, in larger systems, this heat exchanger could be used as a pre-heater for combustion air or as an economizer.

(5) *Combination Systems*: In many cases, particularly with dry injection, it is necessary to combine one or more systems to achieve the desired gas temperature without incurring moisture or corrosion problems.

7.3.2 REAGENT INJECTION

Reagent(s) are injected to remove and/or aid in the removal of the pollutants that are not in the solid phase; specifically gaseous toxics and acid gases. In order to remove these pollutants in the baghouse, they must first be converted from a gas phase to a solid. Although the gas cooling system performs this function on some of the heavy metals, and dioxin/furans, a proper reagent can enhance this removal. Activated carbon has been proven as a very effective adsorbent of mercury, dioxins, furans, and other compounds. See Table 7.3 for a listing of some alternative reagents.

While condensation and adsorption are the primary conversion forces for the toxic gases, a chemical reaction must occur to convert the acid gases to solids. Both the spray dryer and dry injection systems accomplish this in a similar fashion, but the specific acids to be scrubbed, removal efficiency required, and cost considerations will dictate which method is preferable.

Both spray dryers and dry injection systems share the same basic chemical considerations; there must be a sufficient amount of the reagent, it must be well mixed with the flue gas, the temperature must be acceptable, and there must be a sufficient reaction time allowed during the contact period. The designs of the two methods vary as noted below and in Table 7.4.

(1) *Spray Dryer*: Because the reagent is injected as a slurry and the primary reaction occurs on the wetted surface of the reagent, the reaction temperature is low (at the wet bulb temperature of the gas) and the removal efficiency is enhanced for most all acids. Since the spray dryer is an independent vessel, it is relatively easy to design for

TABLE 7.3. Alternative Reagents.

Reagent	Advantages	Disadvantages
Activated carbon	Effective on toxics Small amount required	Little effect on other pollutants
Lime based materials	Low cost Readily available Easy disposal Low molecular weight	Corrosive Irritating to handle High stoichiometric ratio required Poor SO_2 removal Produces deliquescent ash
Sodium based	High efficiency Easy to handle Low stoichiometric ratio required	High cost Limited sources High molecular weight Difficult disposal

TABLE 7.4. **Reagent Injection Methods.**

Method	Advantages	Disadvantages
Spray dryer	High efficiency Reduces reagent usage	High cost Requires higher maintenance Space requirements Possible water carry-over
Dry injection	Simple Low cost Small space requirement	Lower efficiency Higher reagent usage

good distribution of the reagent across the gas stream, and the reaction time is determined by the size of the vessel (although the reactions continue to occur after the gas and reagents exit the cooler, just like in a dry injection system).

(2) *Dry Injection*: The design of these systems varies more, and is more dependent on the acids to be scrubbed. In almost all systems the flue gas cooling takes place before the reagent injection. The reagent can be injected in the ductwork, or a reaction vessel can be used. The distribution of the reagent varies from being dumped through a hole in the duct to Venturi injection sections. Because the reaction temperature is higher than in a spray dryer, the removal efficiency is usually less, and the system is more dependent on optimizing the contact. For this reason, reagent distribution is very important, as is the reaction that takes place in the filter cake on the bags (see below).

(3) *Reagent Material*: Most reagents that have been used are either calcium (lime) based or sodium based. The optimum material is very site specific and depends on the acids to be scrubbed, whether the system uses spray drying or dry injection, the removal efficiency required, material disposal considerations, and cost factors.

7.3.3 PARTICULATE REMOVAL

The flue gas with the solid pollutants, partially reacted acid gas, and unused reagent flows into the particulate removal device where virtually all solid particulate is collected. See Table 7.5.

Baghouses have been favored over electrostatic precipitators because the accumulation of particulate on the bags forms a "filter cake" that enhances the filtration. Also, the reagent materials in the filter cake have a considerable amount of contact with the flue gas and there is more toxic material adsorption and acid gas scrubbing. Contact between the flue gas and reagent is particularly intimate in this area because of the dense cake and very slow gas movement, as opposed to an ESP where the flue gas flows parallel to the collected particulate.

In a baghouse the build-up of particulate on the bags causes a restriction to flow (increase in pressure drop) and periodically the bags must be cleaned. This can be accomplished by different techniques, the most popular being pulse jet cleaning, where a jet of compressed air is blown down the inside of the bags. The bag inflates rapidly and the particulate on the outside of the bag is thrown off and settles into the hopper where it is removed to disposal.

7.3.4 SUBSYSTEMS

The above discussion focuses on those items that directly affect the environmental performance of the system. In addition, there are other systems that support the total operation:

(1) Reagent storage, metering, and delivery to the point of injection
(2) Collected particulate discharge from the hopper(s) and removal to disposal
(3) An I.D. fan system including a draft control that will respond to changing operating conditions
(4) A ductwork system for connecting the components, complete with isolation valves, by-pass system, etc.

While these subsystems do not directly contribute to the environmental performance of the system, their proper design and operation are critical to the success, including the environmental performance, of the system. A system that is designed for 99.9% removal of a pollutant will emit the same amount of that pollutant in a one hour by-pass as it will in six (6) weeks of normal operation. Obviously, unreliable systems that suffer frequent out-

TABLE 7.5. Particulate Removal Devices.

Type	Advantages	Disadvantages
Fabric filter	High efficiency Filter cake helps scrub acids	High cost Higher pressure drop Bag maintenance
Electrostatic precipitator	Low pressure drop	Lower efficiency High cost Complex operation High power consumption

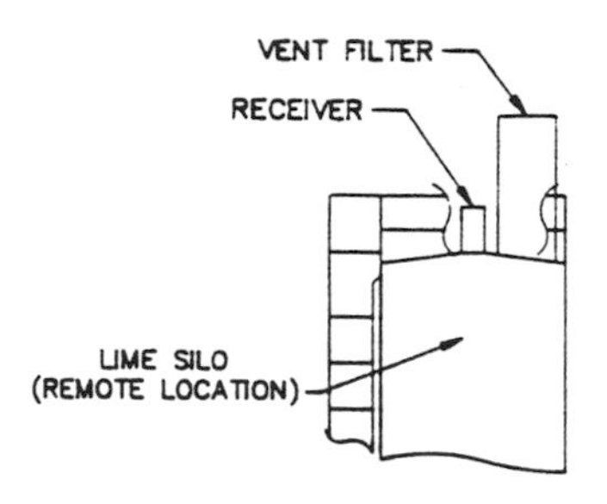

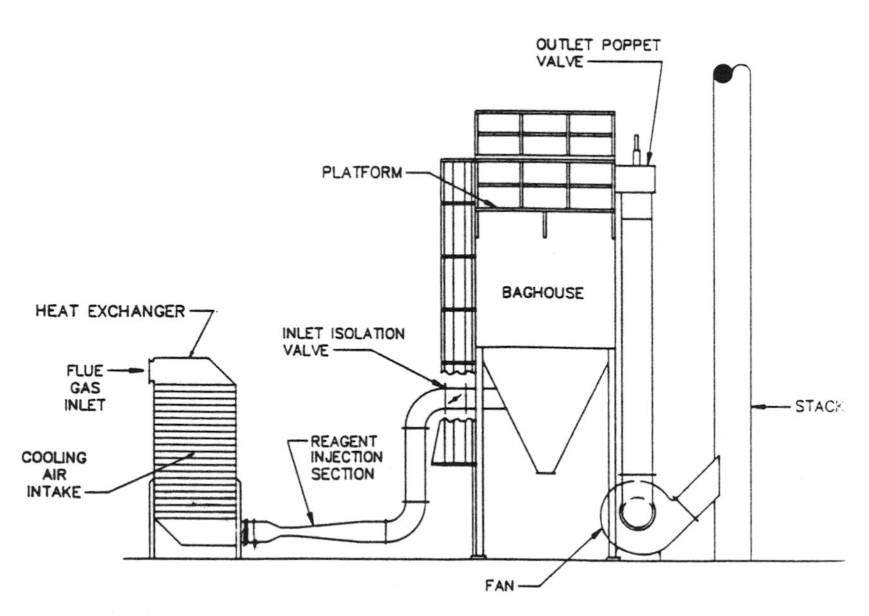

Figure 7.1 Typical general arrangement dry scrubber system with a boiler (not shown) and heat exchanger for termperature control and a dry reagent injection system.

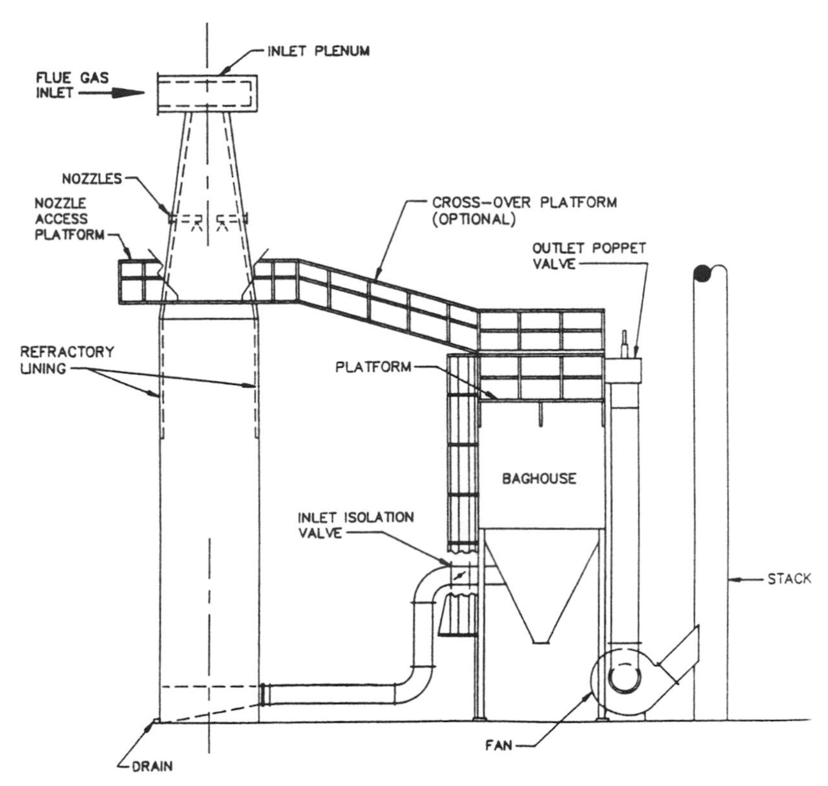

Figure 7.2 Typical general arrangement dry scrubber system with a combination evaporative cooler and spray dryer.

ages provide a far lower level of environmental performance than those that remain on line.

Because these systems have to endure acidic and alkaline corrosion, gas volume and temperature swings, hygroscopic and deliquescent reagents and ashes, etc., on-line reliability and long-term operation without deterioration of performance have been a problem. The design of each of the subsystems is unique to the specific application, and deserves a considerable amount of attention. Two system general arrangements are shown in Figures 7.1 and 7.2.

7.4 SYSTEM PERFORMANCE

The environmental performance of dry scrubbing systems is judged in four basic areas: opacity, particulate, acid gases, and specific compounds. The compounds include the toxic materials, heavy metals, dioxins and furans.

The opacity and particulate performance capability are well established, and scrubbing efficiency for common acid gases has been documented, but there is a very limited data base for the emission of specific compounds (toxics). Since regulations with emission standards for these materials have only recently been enacted, the data base is just now being established. Also, it is very expensive to test for many of the materials, and the systems that have been designed for their removal are new, and the tests that have been performed have produced mixed results, so that performance has not been consistent from system to system. There are so many variables and unknowns that evaluation of operating systems and prediction of future system operation is difficult at this time.

As part of the development of standards and guidelines for medical waste incinerators, EPA surveyed different dry scrubbing systems. The results were published in a draft document in September 1991, and the summary sheet for one test is given as Table 7.6.

These tests and others have generally demonstrated that particulate can easily be reduced to a discharge concentration of 0.015 gr/dscf, corrected to 7% O_2 or 12% CO_2, and in fact, is usually below that level. It follows that opacity will also be low, and is almost always reported at less than 5%, and invisible to an observer. The acid gas removal of these systems has basically been 95%+ for HCl. On medical waste incinerators, the SO_x

TABLE 7.6. Test Data from a Dry Lime Injection/Fabric Filter System (English Units).[a,b]

Pollutant, at 7% O_2	Inlet	Outlet
PM, gr/dscf	0.063	0.002
CO, ppmv	16.9	14.88
2,3,7,8 TCDD, gr/dscf (10^{-9})	0.031	0.035
Total CDD, gr/dscf (10^{-9})	35.1	26.3
2,3,7,8 TCDF, gr/dscf (10^{-9})	3.24	4.00
Total CDF, Gr/dscf (10^{-9})	118	113
Total CDD + CDF, gr/dscf (10^{-9})	153	139
HCl, ppmv	1,739	82
SO_2, ppmv	6.15	5.9
NO_x, ppmv	163	202
Arsenic, gr/dscf (10^{-6})	6.14	5.36
Cadmium, gr/dscf (10^{-6})	244	1.26
Chromium, gr/dscf (10^{-6})	6.32	1.69
Lead, gr/dscf (10^{-6})	1,491	11
Mercury, gr/dscf (10^{-6})	7,130	4,665
Nickel, gr/dscf (10^{-6})	3.92	8.44

[a]Intermittent, controlled-air MWI with a charging capacity of 295 kg/hr (650 lb/hr).

[b]Test condition: G-500 waste at a secondary chamber temperature of 1093°C (2000° F) and a charge rate of 272 kg/hr (600 lb/hr).

levels are so low that it is often not tested, or the concentrations are low enough that percent removal is not a meaningful statistic. However, it has been demonstrated on other applications that a dry injection system will only provide about 50% SO_2 removal, while a spray dryer system can achieve 90% or more. The SO_2 removal efficiency is very temperature sensitive.

The removal of toxic compounds reported in the EPA data does not reflect the improvement that injection of activated carbon provides. The heavy metals that exist as fine solids (such as lead and cadmium) are removed at the same high efficiency as any solid. However, those that had the lowest removal efficiency, or actually showed a negative efficiency, such as dioxins, furans, and mercury are the ones that would have benefited the most by the addition of activated carbon.

7.5 SUMMARY

Dry scrubbing systems are proving their capability to provide control of the various pollutants, including toxics, that are produced by combustion processes, and they are regarded by several environmental authorities as MACT. However, until more systems are in operation and have been tested, there is a lack of a data base that prevents accurate prediction of operating results.

The dry scrubbing name encompasses several different types of air pollution control systems; however, all of these systems are consistent in that they integrate gas cooling, reagent injection and the dry removal of the pollutants, which have been converted to solid particulate.

In part because of the newness of the applications and the competitive nature of the industry, there have been mixed results from a reliability and longevity standpoint. Corrosion, material handling problems, and short bag life have been problems, particularly on the earlier units; however, there is no doubt that these systems, when properly designed and applied, will provide successful and reliable operation and are capable of providing the highest removal efficiencies of the combined pollutants that must be addressed under the new Clean Air Act.

CHAPTER 8

Control of VOCs by Incineration

JOSEPH L. TESSITORE, P.E.[1]

8.1 AIR TOXICS CONTROL

In recent decades, EPA, OSHA, and other government agencies have identified numerous toxic organic compounds. In general, these compounds have been identified based on their potential to pose serious human health hazards. Although many of these compounds are common air pollutants, they are not specifically covered under the National Ambient Air Quality Standards (NAAQS), and only benzene and vinyl chloride are included under EPA's National Emissions Standards for Hazardous Air Pollutants (NESHAP). Many of the remaining compounds were controlled under Volatile Organic Compound (VOC) emission limitations of the NAAQS with the intent of controlling ozone formation.

Since the control of air toxics under the above approach was not totally acceptable from a health risk approach, many states developed regulations and/or control strategies based on emission limitations or Acceptable Ambient Concentration Levels (AACL). This approach created a large variation in the control and enforcement of air toxic emissions across the United States.

More recently, EPA has published a Toxics Release Inventory (TRI) with the intent of identifying the major air toxic pollutants, their emission quantities, and the location and nature of the sources responsible for the emissions. This approach is considered the initial step in the subsequent revision of the Clean Air Act so that emission control strategies may be defined and applied to various industries and/or types of sources.

[1]Vice President, Cross/Tessitore & Associates, P.A., Orlando, Florida.

Many VOCs are "air toxics" or "HAPs" as discussed in Chapter 2. The 25 major air emission chemicals are defined by the EPA TRI. Of these 25 chemicals, 18 are organic compounds, which account for 74.2% of the total air emissions and 85.1% of the fugitive air emissions. These compounds can be further divided into such categories as volatile (easily evaporated) semi-volatile, halogenated, and non-halogenated. Table 8.2 identifies the compounds with respect to the above classifications and presents important chemical and physical characteristics. Table 8.1 also shows that the main fugitive emissions are due to 1,1,1-trichloroethane, toluene, and acetone, which are widely used and highly volatile.

Since the 1990 CAA Amendments, control limits for these are regulated under:

(1) MACT (EPA)
 –high efficiency control
 –enclosure
(2) Ambient air limitations/risk assessment approach (EPA)

TABLE 8.1. Categorization of Waste Gas Streams.

Category	Waste Gas Composition	Auxiliaries and Other Requirements
1	Mixture of VOC, air and inert gas with >16% O_2 and a VOC content <25% LEL (i.e., heat content <13 Btu/ft^3)	Auxiliary fuel is required. No auxiliary air is required.
2	Mixture of VOC, air and inert gas with >16% O_2 and a VOC content between 25 and 50% LEL (i.e., heat content between 13 to 26 Btu/ft^3)	Dilution air is required to lower the heat content to <13 Btu/ft^3. (Alternative to dilution air is installation of LEL monitors.)
3	Mixture of VOC, air and inert gas with <16% O_2.	Treat this waste stream the same as categories 1 and 2, except augment the portions of the waste gas used for fuel burning with outside air to bring its O_2 content to above 16 percent.
4	Mixture of VOC and inert gas with zero to negligible amount of O_2 (air) and <100 Btu/scf heat content.	Oxidize it directly with a sufficient amount of air.
5	Mixture of VOC and inert gas with zero to negligible amount of O_2 (air) and >100 Btu/scf heat.	Premix and use it as a fuel.
6	Mixture of VOC and inert gas with zero to negligible amount of O_2 and heat content insufficient to raise the waste gas to the combustion temperature.	Auxiliary fuel and combustion air for both the waste gas VOC and fuel are required.

TABLE 8.2. Some of the Most Common Compounds Measured as PIC's during the Thermal Destruction of Hazardous Waste.

Volatile Compounds	Semi-Volatile Compounds
Benzene	Naphthalene
Toluene	Phenol
Carbon tetrachloride	Bis(2-ethylhexyl)phthalate
Chloroform	Diethylphthalate
Methylene chloride	Butylbenzylphthalate
Trichloroethylene	Dibutylphthalate
Tetrachloroethylene	
1, 1, 1-Trichloroethane	
Chlorobenzene	

–process feed screening
–process emissions screening
–risk assessment for carcinogenic compounds
–"no threat" level for non-carcinogenic compounds

(3) State regulations
–acceptable ambient levels
–emission limits

In order to effectively control these materials one should first:

(1) Define compound physical and thermal characteristics
(2) Establish thermal destruction criteria
–time and temperature criteria
(3) Volume and nature of waste gas stream
–temperature, CFM, % moisture
–% O_2
(4) Important criteria
–upper and lower HAP explosive limit
–oxygen requirements for complete destruction

Control technologies are noted in Chapter 2 (Table 2.3). Performance range for these technologies is shown in Figure 8.1.

8.2 FUME INCINERATION SYSTEMS

The previous section of this chapter shows that within American industry there is an increasing need to meet requirements for the control of volatile organic compound (VOC) emissions and Environmental Protection Agency (EPA) clean-air standards. Unlike the problem of dealing with particulate-

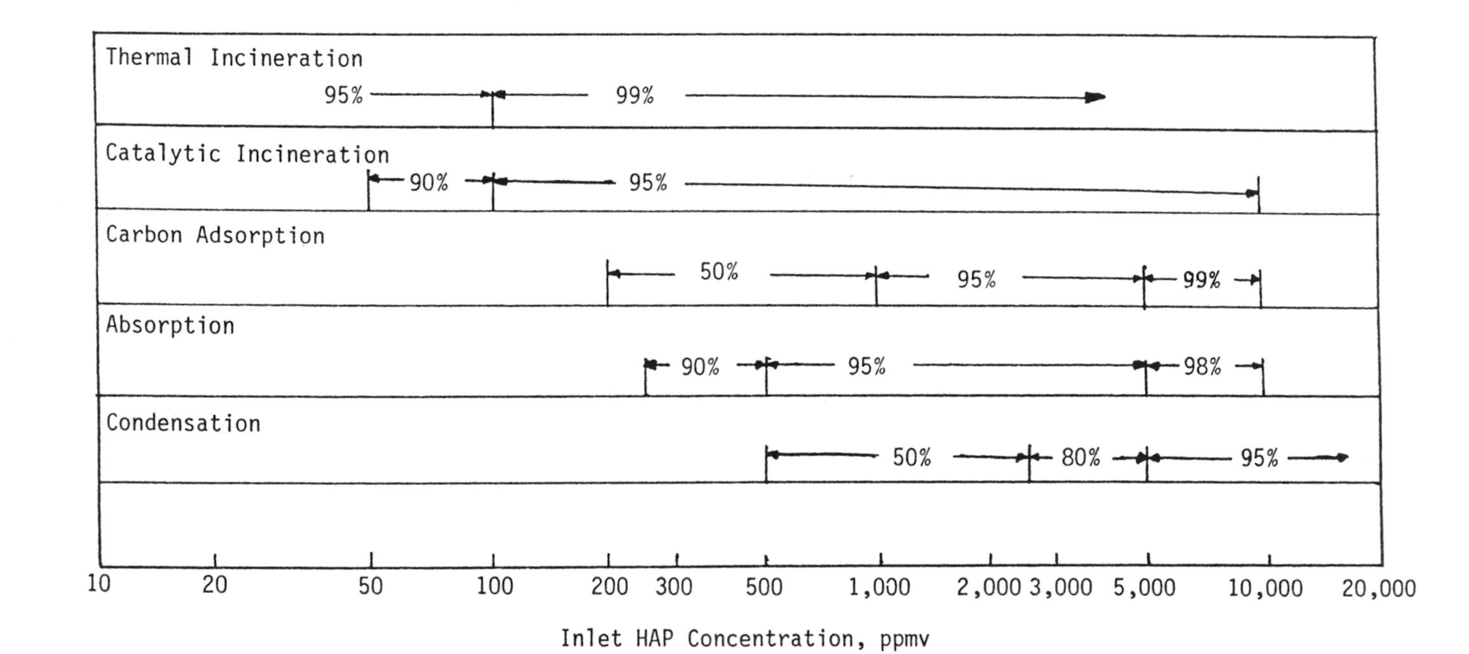

Figure 8.1 Efficiency performance range for control procedures.

laden emissions—a relatively simple process, with minimal operating costs—meeting EPA standards for fume cleanup involves more sophisticated options. Most of these require large capital investments and operating costs. Consequently, the choice of the proper fume incineration system can have a significant impact on profit. The appropriate system can pay for itself within a few years, and lower manufacturing costs thereafter, in contrast with less efficient systems available.

Essentially, there are four types of fume incineration systems, all of which incorporate thermal oxidizers to destroy organic-laden fumes to form carbon dioxide and water vapor, both of which are harmless in the atmosphere. While all the various systems can achieve the necessary cleanup efficiencies, each possesses its own advantages and disadvantages. The purpose of this discussion is to illustrate the differences between fume incineration systems and provide some insight into the most beneficial system for a particular industrial process. The following is a discussion of the systems depicted in Figure 8.2.

(1) *Direct-flame thermal oxidizer (common afterburner)*: This system requires temperatures from about 1200°F up to actual flame temperatures, thereby destroying the VOCs, and maintains them at those temperatures for a given retention time (typically 0.3 seconds and higher). In this process the organics are oxidized through direct contact with the flame.

This system is relatively inexpensive to purchase and install by industrial standards. However, it is the least efficient method of fume incineration due to its high fuel consumption, typically consuming 2000% more fuel than other systems.

It is typically suitable for low-throughput applications. However, its most common use is in the disposal of solid or liquid wastes.

Thermal incineration can occur in a heat exchanger without the presence of a visible flame. This type system presents an option in the type of primary heat exchange method used.

8.2.1 METHOD 1

(2) *Recuperative thermal oxidizer*: This system oxidizes the fumes in a combustion chamber. Unlike the direct flame system, it takes advantage of the existing high temperature gas by passing it through the low-temperature inlet gas stream via an indirect shell-and-tube heat exchanger. This technique allows the incoming VOC-laden gas to be preheated to within 65–80% of oxidation temperatures.

This system is economical to operate when the heat release from the VOCs is sufficient to replace the fuel needed for combustion. Factors

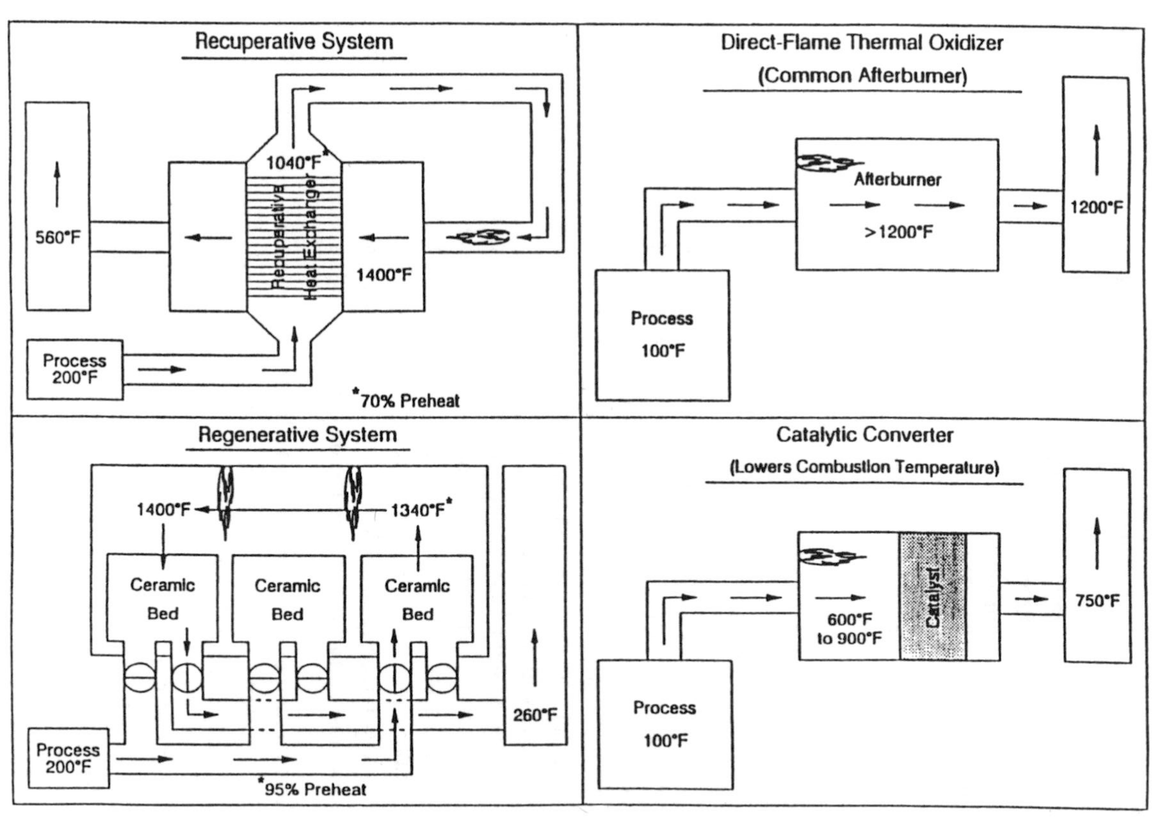

Figure 8.2 Air toxic incineration systems.

such as equipment size, heat transfer coefficients, and stress limit the systems preheat efficiency. Furthermore, the addition of the primary heat recovery system increases the initial capital investment.

The recuperative thermal oxidizer is appropriate for processes that need to incinerate fumes with high solvent contents, typically fumes with an air-to-solvent ratio of greater than 25% lower explosive limit (LEL is the concentration of an organic material that will produce temperatures high enough to sustain flame reactions that depend on forming extremely reactive free radicals).

8.2.2 METHOD 2

(3) *Regenerative thermal oxidizer*: This system utilizes oxidized gases from the combustion chamber by passing it through millions of inert ceramic elements contained in a porous heat transfer section. A system of valves is used to control the inlet of contaminated air into the heat transfer section where the fumes are preheated by convection of the stored heat to within 5% of oxidation temperature on their way to the combustion chamber (up to 95% of the heat of combustion is stored by conduction in the ceramic elements). An uninterrupted flow of VOC-laden gas is processed through the system at all times. This flow is provided for the system by a continuous cycle of alternately storing and releasing heat within three heat transfer sections.

This type regenerative system is simple and reliable, requiring little or no additional fuel, even when fume hydrocarbon levels approach zero. When these levels do approach zero, only 5% of the normally required fuel is needed for oxidation. When coupled with 95% preheat efficiency a pilot burner will provide adequate oxidation temperatures (in most cases). By utilizing a single-speed-reducer mechanical valving system most of the maintenance, common of other valve systems, is eliminated. Regenerative thermal oxidizers are most commonly used for larger process flows with low solvent concentrations, because the high primary heat recovery will reduce operating fuel costs. A 5% LEL or greater will result in almost no fuel usage.

(4) *Catalytic converter*: Catalytic incineration systems require the presence of a catalyst, usually a noble metal, to speed up the oxidation reaction. In this system, the VOC-laden fumes must first be preheated to temperature ranging from 600–900°F. The fumes are then passed through another section of the system containing the catalyst, which is able to thermally oxidize the hydrocarbons present at this reduced temperature range (depending on the type fuel used).

The catalytic converter is more efficient than an after-burner, and is also a simple system. However, it does pose several drawbacks. It performs poorly with fumes contaminated with particulates, resin, heavy metals or silicone, which are commonly found in oven or dryer processes. This is due to the catalyst becoming coated or poisoned with oxidized ash that weakens its ability to oxidize the VOCs at the preheat temperature and may also reduce flow through the system. Thus, additional fuel must be burned to elevate the fumes to high enough temperatures to maintain oxidation. Losses of the catalytic elements must also be minimized because most of the noble metal catalysts are both hazardous and expensive.

This system works well with low-solvent concentration streams. It is best suited for clean ink processes that involve no silicone and therefore, appear very limited in its applicability.

Once the proper thermal oxidation system that best fits your needs has been determined, many parameters are involved in designing its most efficient operation. To ensure the required destruction and removal efficiencies are met, the following areas must be considered in the system design: sufficient retention time for complete oxidation (which also depends on the solvent concentrations in the process fumes); adequate temperature to ensure complete oxidation within that designed retention time; and enough mixing so that all the VOCs come in contact with the oxygen at the above time and temperature provided. These parameters are known as the 3 Ts of the combustion process: Time, Temperature and Turbulence.

It is also required that your system be designed to capture a specified quantity of hydrocarbons being generated within your facility and to yield a specified overall rate of capture/destroy efficiency. This rate depends on a number of variables, such as the nature of the process, type of VOCs emitted, annual production amounts, and geographic location. To determine what the regulations are for your specific process, manufacturers must check with their regional office of the state air-quality board.

8.3 THERMAL DESTRUCTION OF ORGANIC COMPOUNDS

Thermal destruction of organic compounds can occur either in a direct flame incinerator or a catalytic incinerator. Direct flame incineration is a method by which organic compounds in waste gases are directly combusted with or without the use of an additional fuel. In most cases of waste gases, the toxic compounds are too low in concentration to provide a combustible mixture, therefore additional fuel is usually required.

The waste gas stream should be characterized before thermal destruction should be attempted. Table 8.1 lists six categories of waste gas streams with incineration requirements appropriate to each.

In the case of a catalytic incineration, a catalyst accelerates the rate of chemical reaction (destruction) of organic compounds. This results in destruction of organic compounds at lower temperatures and less retention time than required in a direct flame incinerator. In general, a catalytic incinerator will have lower fuel consumption, but will have a higher capital cost due to the presence of a catalyst.

In any oxidation process, time, temperature, turbulence, and oxygen availability are extremely important in achieving the desired destruction efficiency of toxic organic compounds. The reaction kinetics for an organic compound determines the time, temperature, and oxygen requirements. The time, temperature, and oxygen requirements determine the physical size of the combustor and the auxiliary fuel required, which are the most important design considerations for the thermal oxidizers.

8.3.1 TIME-TEMPERATURE RELATIONSHIPS

The thermal destruction of organic compounds is complicated and may involve a series of decomposition, polymerization, and free radical reactions. Most recent studies by the University of Dayton Research Institute and Union Carbide have shown that the most important organic destruction processes are oxidation and thermolytic cracking (pyrolysis). In general, the above studies have involved the development of Thermal Destruction Profiles (TDPs) for numerous organic compounds.

The logarithmic relationship between thermal destruction and retention time for a given temperature can be defined by the following first order kinetic reaction.

$$\ln \frac{C_A}{C_{AO}} = -kt_r \qquad (1)$$

where:

C_A = the concentration of hazardous constituent A at time tr
C_{AO} = the initial concentration of hazardous constituent A
k = the reaction rate constant
t_r = the time required to reach final concentration of C_A

The temperature dependence is included in the equation for the rate constant, k. This may be expressed by the Arrhenius equation:

$$k = A \exp(-E/RT) \qquad (2)$$

where:

A = Arrhenius pre-exponent frequency factor (S^{-1})
E = energy of activation (kcal/mole)
R = universal gas constant (1.987 cal/mol°K)
T = absolute temperature (°K)

Using the laboratory experimental data, A and E values can be determined for each compound of interest. The A and E values can then be used to determine compound destruction characteristics for various temperatures and retention times using Equations (1) and (2).

Based on these A and E values, temperatures were determined for 90%, 95%, and 99% destruction efficiencies (DE) using retention times of 0.5 seconds, 1.0 seconds, and 2.0 seconds, where destruction efficiency is defined by:

$$DE = \frac{[(Mass)_{in} - (Mass)_{out}]}{(Mass)_{in}} \times 100\% \quad (3)$$

8.3.2 PRODUCTS OF INCOMPLETE COMBUSTION

As previously discussed, time and temperature play the major role in the thermal destruction of organic compounds. Unfortunately, the successful destruction of an organic compound or compounds may result in the introduction and/or existence of other organic compounds. These compounds have been identified as products of incomplete combustion (PICs) and have been measured by thermal destruction testing facilities. Table 8.2 lists the most common compounds measured as PICs during the destruction of organic compounds.

A study by the US EPA has identified the existence of PICs in the combustion of organic compounds due to the three following mechanisms:

(1) Compounds which were originally present in the waste feed to the incinerator which were not identified
(2) Compounds which were introduced from other sources other than the waste
(3) Compounds which are actual combustion byproducts

In general, PICs, due to the first mechanism above, are a small percentage of the total PICs observed. For the case of the second mechanism, most PICs can be introduced through the use of auxiliary fuel. This is especially true in cases where fuel oil is used in auxiliary burners, and the fuel oil may contain other waste and/or potential PICs. Therefore, the use of fuel oil as an auxiliary fuel requires the identification of all possible PICs prior to thermal destruction of other organic compounds.

The major source of PICs is due to mechanism 3 (actual combustion byproducts). US EPA studies identified three categories of combustion byproducts.

(1) Identifiable fragments of original constituents which result from partial oxidation or simple substitution reactions. A good example would be the appearance of hexachlorobenzene as a byproduct of PCB combustion.
(2) Reaction products that are the result of complex recombination or substitution reactions. This includes compounds of high molecular weight which may have resulted from complex reactions such as naphthalene, fluoranthane, and pyrene.
(3) Simple fragments that appear as universal byproducts of the combustion of organic compounds. This category includes many low molecular weight compounds such as chlorinated methanes, chlorinated ethanes, and ethylenes. Some of the most common compounds detected are chloroform, carbon tetrachloride, trichloroethylene, tetrachloroethylene, benzene, phenol, toluene, and chlorobenzene.

Most of the testing data shows that PICs can be significantly reduced by increasing temperature, retention time, and oxygen availability. The affect of oxygen availability on the destruction of organic compounds and the PICs will be discussed in the following section.

8.3.3 OXYGEN REQUIREMENT

The thermal stability or destruction of organic compounds is strongly dependent on the availability of oxygen in the thermal environment. Laboratory and subscale testing has demonstrated this dependency for many compounds. The data show that the thermal destruction of a compound increases dramatically with an increase in oxygen availability.

Additional studies conducted at the University of Dayton show TDPs of five organic compounds under excess oxygen conditions, stoichiometric oxygen conditions, and pyrolysis conditions. The results of this study are summarized in Table 8.3. In general, the data show that some compounds (chlorobenzene, trichloroethylene, and toluene) show a large variation in thermal destructibility with oxygen availability while other compounds (freon 113 and carbon tetrachloride) exhibit limited change in TDP with oxygen availability. The data also show that increased oxygen availability yields a higher thermal DE for a given temperature and retention time.

TABLE 8.3. Summary of Thermal Decomposition Tests with Varying Oxygen Availability.

	Temperature Required for 99% Destruction*(°C)				
	Waste Mixture			Pure Compound	
Organic Compound	Excess Oxygen (1,600%)	Stoichio-metric Oxygen	Pyrolysis	Excess Oxygen	Stoichio-metric Oxygen
Freon 113	770	780	780	780	780
Carbon tetrachloride	670	680	680	750	750
Trichloroethylene	730	780	920	780	800
Chlorobenzene	730	800	>1,000	700	900
Toluene	670	750	820	680	820

*Gas phase residence time = 2 seconds.

Other data from the University of Dayton testing show that the number of PICs detected also varies with the oxygen availability during thermal destruction. The majority of the PICs appeared to be the result of pyrolysis type reactions. Also significant levels of PICs were observed at temperatures of 1000°C under pyrolysis conditions while no PICs were observed above 900°C for stoichiometric conditions and above 850°C for excess oxygen availability leads to a more favorable thermal destruction profile with minimum PIC formation.

In addition to these requirements for the destruction of the original organic compound, consideration must be given to minimize the products of incomplete combustion and to control the emissions from the combustion process. These emissions may include particulate, metal fumes, and/or acid gases. In most cases, particulate and metal fumes from gaseous organics are insignificant, but acid gases may be significant due to the chlorine, sulfur and/or fluorine concentrations in the waste gas stream.

Considering the above, the destruction of various organic compounds have been organized to selected groups with thermal destruction criteria being defined for each group.

The compounds in Group 1 (shown in Table 8.4) are non-halogenated organic compounds that can readily be destroyed at 1800°F for nominal retention times of 0.5 seconds. These compounds do not require a scrubber for combustion gas treatment. It should be noted that Table 8.4 gives a retention time range from 0.5 to 1.0 seconds. This range includes a destruction efficiency range from 90% to 99%.

The compounds in Group 2 (Table 8.5) are halogenated organic compounds. These compounds differ from Group 1 compounds in the following manner:

(1) The compounds are generally more refractory and would require

TABLE 8.4. Compounds Which Can Be Destroyed at 1800°F or Greater with Retention Times from 0.5 Seconds to 1.0 Seconds* (no scrubber is required for the combustion of these compounds).

TRI Compounds	
Toluene	Ethylene
Methanol	*N*-butyl-alcohol
Acetone	Benzene
Methyl ethyl ketone	Methyl isobutyl ketone
Propylene	
Other Compounds	
Kerosene	Kerosene
Mineral spirits	Heptane
Formaldehyde	Hexane
Ethanol	

*Higher retention time required for higher destruction efficiency and minimization of PICs.

TABLE 8.5. Group 2 Chlorinated Organic Compounds That Can Be Destroyed at 1800°F or Greater with Retention Times From 0.75 Seconds to 2.0 Seconds* (caustic scrubbing may be required because of HCl formation in combustion products).

TRI Compounds	
1,1,1-Trichloroethane	Trichloroethylene
Dichloromethane	Chloroform
Freon 113	
Other Compounds	
Chlorobenzene	Ethyl chloride
Tetrachloroethane	Trichlorobenzene
Dichlorobenzene	Ethylene dichloride
Perchloroethylene	Methyl chloride

*Higher retention time required for higher destruction efficiency and minimization of PIC's.

TABLE 8.6. Group 3 Refractory Organic Compounds Which May Require 1800°F to 2000°F with Retention Times of 1.0 Seconds to 2.0 Seconds for Satisfactory Destruction (caustic scrubbing may be required for chlorinated compounds).

TRI Compounds	
Tetrachloroethylene	Styrene
Glycol ethers	
Other Compounds	
Carbon tetrachloride	Polyvinyl chloride

longer retention times to achieve acceptable destruction efficiency (>99%).

(2) The compounds contain chlorine and may generate significant quantities of HCl, and therefore the exhaust gases may require treatment with a caustic scrubber.

(3) The existence of chlorinated compounds may also yield higher concentrations of PICs.

The compounds in Group 3 (Table 8.6) are considered the most refractory of the three groups and would require higher temperatures (−2000°F) and longer retention times (2.0 seconds) to ensure good destruction efficiency >(99%) with minimization of PICs.

CHAPTER 9

The Use of Turnkey Contractors and Contract Negotiations

RICHARD P. BUNDY, P.E.[1]

9.1 INTRODUCTION

THE Clean Air Act of 1990 has so changed the nature of air pollution control projects that it should cause a review of the contract structure and conditions that many companies employ.

The Clean Air Act has added significant new complexities to air pollution control projects. We can characterize the conditions before and after the act as follows.

9.1.1 PRE-CLEAN AIR ACT OF 1990

REGULATIONS: Well-established and clear lines of authority.

PERMITTING: Precedents had been set, the permitting program was known, and results were fairly predictable.

POLLUTANTS: In most cases the only pollutants addressed were overall particulate emissions and selected acid gases.

EQUIPMENT: On most applications the most cost effective equipment design and its performance capability had been determined (i.e., wet scrubber, baghouse, electrostatic precipitator) and the only decision regarded the specific details of design and construction. Usually only a single piece of equipment was required in the air pollution control train.

TESTING: The testing methods had been well established, in use for years,

[1]President, Bundy Environmental Technology, Inc., Reynoldsburg, Ohio.

and there were only a limited number of pollutants to be tested for. There was little opportunity for error.

STACK MONITORING: In many cases, no continuous stack monitoring was required. As a maximum, emission monitoring was only required for SO_2 and opacity.

9.1.2 POST-CLEAN AIR ACT OF 1990

REGULATIONS: The provisions of the Clean Air Act have not been reduced from the Congressional mandates to practical performance standards on most specific applications. Many individual states are ahead of the Federal EPA in addressing the new requirements. However, the users will have to meet the provisions of the Federal standards if they are more stringent, once they are established. This adds a serious unknown element to any project.

PERMITTING: Since the regulations are fairly new, there are few if any precedents established. There is more emphasis on boundary-level concentrations than stack emissions, so modeling becomes a major activity. This can create almost a trial and error design technique in some cases. A design is selected and modeled, based on stack emissions. Then, if this does not produce results that are acceptable to the permitting agency, the system is redesigned with new parameters.

POLLUTANTS: Almost every combustion application will be regulated for particulate, acid gases (primarily sulfuric and hydrochloric), heavy metals, dioxins and furans, hydrocarbons, etc. These can all be regulated both as stack emissions and ambient air concentrations.

EQUIPMENT: Pollution control systems will typically involve multiple pieces of equipment to handle the multiple pollutants. Since there is such a limited data base of equipment performances and alternative costs, it is likely that more than one control technology will have to be considered.

TESTING: Testing for multiple pollutants and at minute concentration levels is far more of a technical challenge and financial burden. EPA test methods are changing and each permit application involves establishing a new test protocol. The opportunity for error is considerable and the cost impact substantial.

STACK MONITORING: The specific CEM requirements are likely to be established on a case-by-case basis, but may include opacity and SO_2 as before, plus CO, O_2, temperatures at two or three points in the system, and more.

In the Pre-CAA of 1990 era, the design, procurement, and installation of an air pollution control system was primarily an exercise in system layout optimization, physical construction specifications, and cost minimization.

In those cases many owners were very comfortable in purchasing the components, contracting for the permitting work with an engineer, and overseeing the installation.

The far more complex situation that exists now requires re-evaluation in contracting procedures. Fiscal impact and technical issues are going to develop during the course of the project that could not have been well defined initially.

Many companies that had the technical and management resources to handle a project from permitting through installation in the past now find that they need the support of a turn-key contractor. Others will still want to manage the project themselves but need to give considerable thought to the structure and content of the contracts that they establish with the environmental consultant, design engineers, equipment suppliers, installation contractors and testing company that will all be a part of their environmental project.

9.2 PROJECT ELEMENTS

Although each project is unique, following is a program that most will follow to a reasonable extent.

(1) Definition of the project scope and operating parameters
(2) Preliminary definition of the environmental regulations
(3) Review of the alternative technologies and selection of the one(s) to be considered. This will typically involve:
 –selection of a preliminary approved vendor list
 –budget bidding
 –review of the project with the assumptions made in steps #1 and #2
 –elimination of some alternative technologies and/or bidders
(4) Competitive firm bidding
(5) Bid evaluation and preliminary selection of a vendor. At this point, the selected vendor might only be given a limited contract, pending permitting
(6) Initial permit application
(7) Regulators questions, response by the owner with input from the engineer and vendor. During this period there may be system design changes and/or additions.
(8) Receipt of the permit to construct
(9) Review primary equipment supply contracts and make any modifications required. Finalize construction contracts.
(10) Contract for CEM system and emission testing

(11) Equipment received
(12) Installation of system
(13) Start-up and initial operation of the equipment
(14) Performance testing of the equipment
(15) Operating permit received

There are obviously other programs such as funding the project, negotiation of back charges and changes, etc., that take place on parallel paths.

9.3 CONTRACT MANAGEMENT ALTERNATIVES

There are different alternative ways to handle the project during the fifteen steps, with the primary difference being the amount of control and involvement that the owner wants to have versus contracting the work.

9.3.1 TURNKEY CONTRACTOR VS. INDIVIDUAL CONTRACTS

For the purposes of this discussion, we need to clarify the definition of "Turnkey contractor." This is sometimes confused with purchasing equipment on an installed basis.

A turnkey contractor has total responsibility for the project within well-defined limits and has to accept the liabilities within those limits. Although turnkey usually means the contractor is responsible for the total performance until the key is literally turned on and the system is fully operational, permitted, and functional, the limits do not necessarily have to be the entire project.

An equipment supplier who is installing his equipment has a much more limited scope and typically has exclusions for anything that changes or anything that he was not aware of initially. He also accepts responsibilities and liabilities within defined limits; however they are usually described in more specific terms. Since there are several suppliers on a project and each has its own defined scope, it is the responsibility of the owner or the turnkey contractor to assure that everything that is required is taken care of, and that all the individual defined responsibilities add up to a working system that complies with the overall project goals.

The most important scope related aspect of a turnkey contract is the list of exclusions, or "purchaser-to-furnish" items. The contractor may rely on the owner to perform certain functions such as bringing utilities to the area, disposal of collected pollutants, definition of certain operating parameters, etc., but if he has not stated otherwise, the contractor is responsible for the work.

On the other hand in a supply contract (even if it is on an installed basis) the more important part of contract relating to scope is the listing of what is included. Although equipment suppliers will often define what is not included in a "purchaser-to-furnish" category, this listing is generally more for the convenience of the customer. It is informative rather than providing for complete assurance that everything in the project is covered. Usually the last item on the purchaser-to-furnish list is something like "all equipment not specifically described in this proposal." In other words if the vendor has not specifically stated that he is supplying an item then he is not.

There are three (3) basic reasons to consider turning a project over to a turnkey contractor:

(1) Better cost definition: Firm or well-defined bids are solicited and the owner has the comfort of knowing his costs. The contractor accepts the cost estimating risks.

(2) Management time: The owner simply may not have personnel with the time to manage the project.

(3) Technical capability: Air pollution control projects for many companies are normally a one-time affair, and the owner may not employ people with the experience and technical capability to manage the project.

On the other hand, there are two (2) principle disadvantages to subcontracting the entire project to a turnkey contractor:

(1) Higher costs: A turnkey contractor obviously deserves a profit for his work and acceptance of risk.

(2) Loss of control: The turnkey contractor will make decisions based on his obligations under the contract and does not have the same priorities as the owner.

There can be compromises in the contracting structure that allow the owner to achieve some of the benefits of a turnkey contract and minimize the disadvantages.

There is a distinct difference in the activities in the first eight (8) steps above as opposed to the last seven (7). During the first eight, the project requirements are not firmly defined. If a contract is written for a company to do this portion of the work, it is not reasonable to expect to be able to do so at a firm price, unless the contractor includes considerable contingencies.

Also, the first eight steps do not involve substantial cost, yet their results dictate the total project requirements. In many cases the owner could maintain control and minimize costs, yet still be able to off-load management time and have known costs by contracting with a consulting engineer to do the environmental and preliminary design work initially (steps 1 through 8) then make a decision regarding turnkey or component contracting for the actual system procurement and installation.

9.4 ELEMENTS OF A CONTRACT

Regardless of whether the contract is for environmental and design engineering services, equipment supply, field construction, or a complete turnkey project, there are common basic elements of the contract. Their relative importance and specific purpose will vary depending on the type of contract, but it is worth a review of these elements in all cases.

(1) Definition of the operation: In all cases, the owner is going to have to provide some basic definition of his operation. The level and detail of the definition will vary depending on the point in the program.

–To an environmental engineer involved at the outset, the operation may be described for example, as briefly as "a municipality that needs to incinerate X tons of waste per month." The engineer then would help refine this definition for the equipment vendors.

–A turnkey contractor or major equipment supplier for this MSW incinerator project would expect to be told the burn rate per hour, the heating value of the waste, the number of incinerator trains, etc.

–The air pollution control bidder would like to see this defined even more tightly as to the flue gas produced. He would like to have the gas volume, temperature, certain flue gas components (H_2O, O_2, CO_2, SO_2, and HCl), and the loadings of all the pollutants that must be controlled (particulate, heavy metals, dioxin, etc.). As discussed below, it may not be possible (or prudent) to establish this level of definition.

(2) Description of the goals of the project: At the most basic level, the goal is to receive and maintain an operating permit. As the project progresses, the goals will become much more specific. They will be reduced to exact levels of performance in regard to each controlled pollutant, and these will become the points requiring performance guarantees.

In many cases, the performance level of the equipment and the definition of the operating conditions becomes one of the more significant points of negotiation. It generally favors the owner to define the operation in the broadest sense (as described in 1 above) while the air pollution control system vendor wants to know the specific input level and guaranteed outlet level of each pollutant (as described above).

(3) Establish the terminal limits of the contract: Whether the contract is being negotiated with a turnkey contractor, an engineer, or component vendors, somewhere the limits of supply and responsibility must be

established. As described above, it is typical to tell a turnkey contractor what *not to* include and tell any sub-vendor what *to* include.

(4) Provide specifications for component design: Depending on the structure of the contract, the component design specifications may run from "fit for the purpose intended" (which no vendor is likely to accept), to "suitable for ____ years operating service" (which is a very reasonable definition, but is difficult to enforce), to "make it out of ____ material."

The latter case is the easiest way to be able to compare bidders and is probably the most reasonable approach, so long as the specification is prepared by someone knowledgeable. If the specification is not correct, there is little or no recourse to the supplier.

(5) Describe contract conditions: This is the so-called "boilerplate" of the contract. Details of this portion of the contract are usually very project-specific, and will not be discussed here.

(6) Price and payment terms: Naturally, this is one of the most significant parts of the contract, but the interests and goals of each party are obvious and will not be discussed here.

(7) Guarantees and remedies: The guarantees that are provided to assure compliance with the goals of the project are obviously very important. What happens in the event the guarantee is not met is of even greater importance. This is an area where the implications of the Clean Air Act are felt especially hard.

In the past, companies with any level of experience knew the capabilities of their products. The performance levels had been established based on the performance of similar systems and were not particularly demanding. If a system failed to meet the guaranteed performance, it was more likely a case of a material defect that could be repaired (leaking weld, bad bag, low water pressure, etc.), or the vendor tried to cut corners a bit too much (high air-to-cloth ratio, poor bag selection, low wet scrubber pressure drop, undersized electrostatic precipitator) and would have to pay the price to upgrade the system.

The requirement for regulation of toxic materials is so new that there is a very limited data base available for both the regulators to establish limits and the equipment suppliers to predict results. Therefore, there is a greater risk of failure.

When this happens, there is a dilemma. Many things could have happened, and none have a very ready solution.

–The test could have been flawed. Since the testing methods are new and the levels being tested for are infinitesimal, it is possible that errors could have occurred. However, how is that known without retesting (at considerable expense)? Even that will not reconstruct the operating conditions at the time of the original test.

–The inlet conditions may not have been in accordance with the contract. If the owner established an inlet concentration of the pollutant, the only way to know if the actual condition met the specified level would have been an inlet test. Unless regulations required it, it is unlikely that anyone would have spent the money required to test the inlet conditions.

–The system may have simply failed to meet the required performance level. Because of the new codes, stringent regulations, and limited data base, there are going to be misapplications and failed systems.

If a system fails to meet the guaranteed levels, it could be the testing company's error, the owner may not have correctly specified the application, the APC system may be inadequate, or it is even possible that the regulatory agency set a goal that could not be realized. There is no easy, inexpensive method to establish blame, and probably no easy solution to implement, unless it was just the result of a simple error. It is at this point that the skill with which each party negotiated the contract will be apparent.

9.5 SUMMARY

The Clean Air Act of 1990 has added importance to three (3) different contract issues. A company's assessment of the importance of these issues and an evaluation of their abilities to handle them will dictate the structure of their project's management and should receive considerable attention from them as they negotiate the contracts for the project. These three issues are:

(1) *Project Changes*: During the permitting process, applications are being reviewed on a more individual basis and the results are less predictable. Final permit conditions or issues raised during the evaluation may change the design or even the generic type of air pollution control equipment. This, coupled with the lack of assurance of a permit being granted, restricts the owner from fully releasing vendors prior to receipt of a permit.

(2) *Emission Limitations*: Previously emission limits were primarily based on the capability of demonstrated systems and there was a considerable data base to use to predict results. There was little chance of failure.

New codes require the highest possible removal efficiencies of numerous pollutants and there is a very limited data base from which to predict results. There is every likelihood that there will be systems that cannot meet the emission limitations of the permit conditions.

(3) *Guarantee Remedies*: Since prior emission limitations were virtually always achievable, companies that failed emission tests could be expected to correct an error and get their systems into compliance.

A system that exceeds a dioxin limitation that may read something like 5.0 E-10 lbs/hr by 20% or 200% does not have clear alternatives. There are issues of inlet loadings, testing accuracy, who pays the retesting costs, what equipment or operating changes should be made, is there an alternative percent reduction standard that is applicable, etc.

These points and others deserve considerable thought during the contracting stage. However, the structure of the contract should be such that the owner and the vendor consider themselves partners in achieving system success rather than adversaries in establishing blame.

CHAPTER 10

Permitting and Testing – The Needs and Costs

HOWARD E. HESKETH, Ph.D., P.E.[1]

10.1 RATIONALE—AIR QUALITY REGULATIONS

10.1.1 THE CLEAN AIR ACT—THE BASIC DRIVING FORCE

SINCE it was passed in 1955, the Clean Air Act (CAA) has been the basis for regulating emissions of air pollutants to protect human health and the environment. The Clean Air Amendments to the CAA, passed in 1970, allow EPA to delegate responsibility to state and local governments to prevent and control air pollution at its source. Emissions from stationary sources (industrial facilities) and motor vehicles are regulated under the Act. Stationary sources must obtain permits that specify the amount and type of allowable emissions from an air quality regulatory agency. Modifications to an existing facility or a new facility that emit pollutants are subject to the provisions of the Clean Air Act.

10.1.2 CAA AMENDMENTS

The CAA remained virtually unchanged from 1977 until it was significantly amended in 1990. Congress and public interest groups believed the CAA was ready for revision and cited, as evidence of the Act's failure, the 60 metropolitan areas that did not achieve the ozone standard by the CAA deadline of December 31, 1987. While many areas made substantial progress in improving air quality, EPA and Congress pressed these statistical data as reasons to amend the Act.

[1]Southern Illinois University at Carbondale, Carbondale, Illinois.

The House, Senate and EPA made sweeping amendments to the CAA as noted in chapter 2. These amendments cover a variety of issues raising a host of novel approaches to old problems, such as:

- hazardous air pollutants
- acid deposition control
- nonattainment areas
- ambient air quality standards
- vehicle emission standards
- municipal incinerator emissions
- outer continental shelf and marine activities

10.1.3 AIR TOXICS PROGRAMS

New regulations have emerged under the CAA for control of air toxics. Even so, many of these requirements are on the state level. Regulations governing toxic air pollutants have been prompted by a host of driving forces, including:

- government agency concern over potential adverse effects of these pollutants
- public pressure
- environmental impacts from accidental releases of toxic substances

Both the federal EPA and state agencies are in the process of developing administrative rules, defining control programs, and collecting background information on toxic pollutants. Air toxics regulations are slow to develop, reflecting in part the complex nature of air toxics exposure and impacts. These evolving programs, however, will have major implications for the operation of existing facilities and the design of new ones.

10.1.3.1 Federal EPA Air Toxics Programs

Under the CAA, EPA is required to regulate hazardous air pollutants that in its judgment cause adverse health effects. To date, EPA has initially listed 188 chemical pollutants, but is investigating scores of additional chemicals. EPA presently provides, through the *Federal Register*, a formal notice of intent to "list" or "regulate" a given chemical; however, emission limitation standards only apply after EPA issues a final regulation. Often, early informal knowledge of EPA's intentions to regulate a given substance may be obtained by reviewing health assessment documents and other reports. It is through this process that there are opportunities for public comment to EPA on prospective control of potentially hazardous chemicals.

EPA is currently evaluating the potential health risks associated with many toxic air pollutants, including potentially carcinogenic or mutagenic substances and those with suspected chronic effects. The CAA provides that within one year of listing a particular hazardous air pollutant, the EPA must propose an emission limitation for that pollutant under the National Emission Standards for Hazardous Air Pollutants (NESHAP). In practice, EPA's timetable is much delayed. A NESHAP regulation establishes a uniform national emission standard for sources that emit the listed hazardous pollutant. The standard must reflect an "adequate margin of safety" to protect public health. Such risk assessments may involve the generic use of source information, atmospheric dispersion models, and health or cancer risk factors derived from earlier health assessment studies.

EPA has long argued that the NESHAP program may not be the most appropriate means of regulating potentially hazardous pollutants. Rather, EPA believes a coordinated effort with state and local agencies may be more effective in developing regulations for discrete, local impacts.

Air quality issues also arise from federal hazardous waste legislation, specifically:

- Resource Conservation and Recovery Act (RCRA)
- Comprehensive Environmental Response, Compensation and Liability Act (CERCLA or "Superfund")
- Title III of the Superfund Amendments and Reauthorization Act (SARA), known as the Emergency Planning and Community Right-to-Know Act

10.1.4 AIR POLLUTION CONTROL TRENDS

Air pollution control is at the threshold of a new regulatory era. Slow moving federal initiatives to control air toxics have led several states to take a leadership role in adopting air toxics control programs. Some of these programs are designed in response to public concerns over health and safety issues, but the intensity of concern has outstripped scientifically valid determinations of public health risks. A notable example is Proposition 65, which was adopted by California voters in November 1986. Proposition 65 places the burden of proof on business to show that exposures to listed carcinogens do not present a "significant" risk, whether an individual has been harmed or not. Further, the state agency responsible for implementing the statute sets acceptable exposure levels for some listed carcinogens at three or four times below the average ambient concentrations of these substances in the urban environment. Proposition 65 is indicative of the public's skepticism that current environmental laws and regulations are adequate to protect their health.

Ironically, the push at the state level for the control of air toxics may lead to greater control of air toxics on the federal level as well. SARA Title III will provide regulators and the public with a wealth of information on the emissions of air toxics into the environment. These data may be used to drive new federal legislation and regulations designed to further reduce toxic emissions. The proposed Clean Air Act amendments under consideration in the Senate include a major provision to strengthen the federal regulation of air toxics. Existing stationary sources, largely exempt from air toxics regulation so far, may be the focus of the new air toxics initiatives.

Existing sources in certain areas of the country will also be required to modify operations in response to several recent federal regulatory programs designed to control criteria pollutants. EPA's PM-10 primary and secondary ambient air quality standards, promulgated July 1987, are already forcing potential new sources to employ more costly control technologies and, in some cases, to secure expensive air pollution offsets. EPA's proposed nitrogen oxide increment under the federal PSD program also would require costly controls. Visibility impact analyses under state and federal new source review programs will become increasingly rigorous. When EPA finalizes its regional haze regulations, some existing sources may have to apply retrofit control technology to further limit visibility-reducing sulfate and nitrate aerosols.

All of these federal programs follow the regulatory matrix established under the Clean Air Act. The amendment of the Act has hastened the burgeoning post-1987 regulatory phase, with the attendant SIP revisions for nonattainment area. The further control of existing stationary sources in these nonattainment areas will likely force major policy questions to the surface.

Many ozone nonattainment areas have required existing stationary sources to successively lessen their hydrocarbon, and in some instances, nitrogen oxide emissions in an effort to reduce ambient ozone concentrations. If additional control measures are not sufficient to achieve the federal ozone standard, it may force state and local agencies to consider non-traditional control strategies, including industrial throughput limitations or emission caps on stationary sources and alternative fuels for mobile sources.

Agencies may revisit indirect source control strategies that they flirted with in the early 1970s. Indirect controls are transportation control measures to limit vehicle emissions, and may include high occupancy vehicle lanes on freeways or limits on the number of parking spaces in new commercial developments. Chronic nonattainment areas, while driving new, expensive and innovative control measures, will force a fundamental policy question – is the increased cost of controlling ozone precursors worth the potential health benefit?

10.2 PREPAREDNESS, PREVENTION AND CONTINGENCY PLANS

Air Resources and Protection of Community/Global Health and Welfare procedures are detailed in Federal and local permit procedures. Details on air emissions for a described facility and operating procedures must be contained in the permit application to prevent significant air deterioration and the associated health risk deterioration. Permits also contain details of *Preparedness*, *Prevention* and *Contingency* Plans. An example of a PPC plan outline is:

Section	Title
A.	General Description of the Industrial or Commercial Activity
B.	Description of Existing Emergency Response Plans
C.	Organizational Structure for Implementation of the PPC Plan
D.	Material and Waste Inventory
E.	Spill and Leak Prevention and Response
F.	Material Compatibility
G.	Inspection and Monitoring Program
H.	Preventive Maintenance
I.	Housekeeping Program
J.	Security
K.	External Factors
L.	Internal and External Communications or Alarm System
M.	Employee Training Program
N.	List of Emergency Coordinators
O.	Duties and Responsibilities of the Emergency Coordinator
P.	Chain of Command
Q.	List of Agencies to be Notified
R.	Emergency Equipment
S.	Evacuation Plan for Installation Personnel
T.	Emergency Response Contractors
U.	Agreements with Local Emergency Response Agencies and Hospitals
V.	Pollution Incident History
W.	Implementation Schedule

10.3 OTHER NEEDS OF TESTING

In addition to the regulatory items noted and to the protection of health and welfare, there are the obvious reasons for testing such as:

- to obtain equipment sizing and design data

- for permit data, including environmental assessment; air quality modeling; and PSD calculations
- to certify compliance on an annual basis

10.4 COSTS OF PERMITTING, COMPLIANCE STACK TESTING AND CONTINUOUS EMISSION MONITORING (CEM)

10.4.1 PERMITTING

Title V of the Clean Air Act requires tens of thousands of air pollution sources to obtain an operating permit incorporating all applicable requirements under the Act. EPA recently promulgated its controversial Title V regulations, which establish the minimum elements for state permit programs.

The new permit system is among the most important changes made by the 1990 Clean Air Act Amendments, and will significantly alter the way companies comply with air pollution requirements. Previously, the Act only required certain sources to obtain a new source review permit before constructing or modifying the facility (although many states established operating permit systems on their own). Now, all states must adopt operating permit programs consistent with the minimum federal requirements, and submit them to EPA by November 1993. Even though EPA has established minimum requirements, these programs are likely to vary widely from state to state.

The permitting process includes the submission of required paperwork and the development of compliance plans. These plans should be a format to bring a facility into emission compliance and includes O & M procedures for the control equipment and details operator training programs.

These costs include both in-house expenses necessary to prepare the permits and filing fees with local, state and federal regulatory agencies. In-house costs may include stack testing and CEM (discussed below), as well as design, modeling, engineering, public hearings and a vast range of technical and non-technical items. These in-house costs can amount to hundreds of thousands of dollars.

Filing fees for permits vary from one location to another. Table 10.1 contains excerpts from the State of Florida Department of Environmental Regulations (DER) and gives example fee charges for construction, operation and closure of certain systems. Some fees are only $1,000 (or less), while others can be over $30,000 (in 1993). In some states, permit fees run up to $100,000. In addition emission fees currently are about $10 per ton of pollutant per year (range is about $8/ton to $35/ton for HAPs).

These fees vary depending on the proposed system magnitude and type and, therefore, the amount of time necessary to process the information by

TABLE 10.1. **Permit Fee Charge Example Excerpted from Florida Department of Environmental Regulations (DER), dated October 1991.**

DER-17 Subsection #	Part I: General	Fee, $
4.050(4)(a)2.e	e. Operation permit for a minor source required to measure actual emissions by any method other than stack sampling (such as visible emissions observation or continuous emissions monitoring)	$1000
	f. Operation permit for any minor source not required to measure actual emissions	$750
4.050(4)(f)	(f) Solid waste permits	
	1. Construction permit for a class I facility	$10,000
	2. Construction permit for a class II facility	$10,000
	3. Construction permit for a class III facility	$6,000
	4. Construction permit for a waste-to-energy facility not covered by the Electric Power Plant Siting Act	$10,000
	5. Construction permit for other resource recovery facilities	$2,000
	6. Construction permit for an incinerator	$3,000
	15. Operation permit for a waste-to-energy facility not covered by the Electric Power Plan Siting Act	$10,000
	16. Operation permit for other resource recovery facilities	$1,000
	17. Operation permit for an incinerator	$1,000
4.050(4)	(4) Processing fees are as follows:	
	(a) Air pollution source permits	
	1. Construction permits	
	a. Construction permit for a source having potential emissions of 100 or more tons per year of any single pollutant and requiring a prevention of significant deterioration (PSD) or nonattainment area (NAA) new source review permit	$7,500

(continued)

TABLE 10.1. (continued).

DER-17 Subsection #	Part I: General	Fee, $
4.050(4) (continued)	b. Construction permit for a source having potential emissions of 100 or more tons per year of any single pollutant but not requiring a PSD or NAA new source review permit	$5,000
	c. Construction permit for a source having potential emissions of 50 or more tons per year, but less than 100 tons per year, of any single pollutant	$4,500
	d. Construction permit for a source having potential emissions of 25 or more tons per year, but less than 50 tons per year, of any single pollutant	$2,000
	e. Construction permit for a source having potential emissions of 5 or more tons per year, but less than 25 tons per year, of any single pollutant	$1,000
	f. Construction permit for a source having potential emissions of less than 5 tons per year of each pollutant	$250
	2. Operation permits	
	a. Operation permit for a major source required to measure actual emissions by stack sampling	$2,000
	b. Operation for a major source required to measure actual emission by any method other than stack sampling (such as visible emissions observation or continuous emissions monitoring)	$2,000
	c. Operation permit for any major source not required to measure actual emissions	$2,000
	d. Operation permit for a minor source required to measure actual emissions by stack sampling	$1,500
	(h) Hazardous waste permits	
	1. Construction of container and/or tank hazardous waste storage facilities	$15,000
	2. Construction of container and/or tank hazardous waste storage and treatment facilities	$20,000

TABLE 10.1. (continued).

DER-17 Subsection #	Part I: General	Fee, $
4.050(4) (continued)	3. Construction of landfill, surface impoundment, waste pile, land treatment, and miscellaneous unit facilities	$25,000
	4. Construction of hazardous waste storage, treatment and/or disposal facilities with an incinerator for treatment of hazardous wastes generated on-site	$25,000
	5. Construction of commercial treatment, storage, and/or disposal facility with a commercial incinerator managing hazardous wastes generated off-site	$32,500
	6. Operation of container and/or tank hazardous waste storage facilities	$10,000
	14. Closure of hazardous waste storage, treatment and/or disposal facilities with an incinerator for treatment of hazardous wastes generated on-site	$15,000
	15. Closure of commercial treatment, storage, and/or disposal facilities with a commercial incinerator managing hazardous wastes generated off-site	$32,500
	16. Hazardous waste research, development and demonstration facilities	$4,000
	17. All other hazardous waste facility permits or authorizations for which a specific fee is not specified	$32,500
	(q) All fees shall be deposited in the Florida Permit Fee Trust Fund created pursuant to section 403.087(5), F.S.	
	(r) If the department requires by rule or permit condition that any specific permit be renewed more frequently than once every five years, the permit fee shall be prorated based upon the permit fee schedule in effect at the time of permit renewal. Upon issuance of such a permit, a prorated refund of the fee shall be returned to the applicant. This provision does not apply to permits issued for less than five years which could be extended to five years without the filing of an application for renewal. However, applications for permits to continue operation of a facility where an existing permit has or is about to expire in accordance with section 403.087(1), F.S., shall be accompanied by the appropriate processing fee.	

the agency. As an example of how a single system can have over two dozen variations, see Table 10.2 for flue gas desulfurization (FGD) retrofit options. These are estimated using 1992 data from the Electric Power Research Institute (EPRI) for a 300 MW size system with 2.6% sulfur coal 90% SO_2 removal unless otherwise noted. These data are for retrofit systems which are more expensive than new systems.

10.4.2 COMPLIANCE STACK TESTING

Stack testing costs can range considerably, depending on what is required. A "simple" triplicate US EPA Method 5 test for particulates, including flue gas analyses for CO_2, O_2 and CO can be $2,500 or more, plus travel costs of the test team. A source outlet VOC test on an "ideal" situation using Method 25A with report would be expected to cost $3,000 to $6,000, plus travel. More complex testing for volatile organic compounds (VOC) with VOST or similar test trains can be $20,000 to $30,000. As a "rule of thumb," add $3,000 test costs for every pollutant added to the test requirements.

TABLE 10.2. FGD Retrofit System Costs in 1992 Year Dollars.

Example Systems	Capital Cost, $/KW	Operating Cost, $/Ton SO_2 Removal
A. Wet scrubbing processes		
Limestone/forced oxidation	210	540
Limestone/wallboard gypsum	225	550
Limestone/inhibited oxidation	205	530
Limestone/dibasic acid	202	525
Pure air/Mitsubishi	185	470
Magnesium-enhanced lime	185	515
Bischoff	215	550
Saarberg Holter	180	460
Noell/KRC	225	565
Lime dual-alkali	185	505
Limestone dual-alkali	190	480
B. Dry injection processes		
Lime spray drying	150	470
Tampella LIFAC (80%)	215	640
C. Circulating fluid bed		
Lurgi	140	380
D. Sulfur recovery processes		
Wellman-Lord	270	600
SOXAL	230	610
Magnesium oxide	275	615

Remember that once a facility is in compliance, annual certification of compliance must be made. This includes emission testing, control device operation and maintenance, and personnel training.

Some of the US EPA reference methods (40 CFR, Part 60, Appendix A) for VOCs are noted here:

(1) *Method 18* uses techniques in gas chromatography to identify specific gaseous hydrocarbon emissions.

(2) *Method 21* uses an organic vapor analyzer (OVA) to survey process equipment and identify fugitive VOC emissions.

(3) *Method 23* measures dioxins and furans.

(4) *Method 24* determines the percent VOCs and density of the coating, as well as the solvent.

(5) *Method 24A* determines volatile matter content, water content, volume and weight of solids in paint, varnish, lacquer or related surface coatings.

(6) *Method 25* measures VOC as total gaseous nonmethane organics. Emissions are determined in the laboratory using a gas chromatograph.

(7) *Method 25A* measures the total gaseous organic concentration using a portable flame ionization analyzer. This analysis is performed on site.

Some methods used to determine metals and other non-organic emissions are:

(8) *SW846 Method 0030* determines volatile emissions.

(9) *SW846 Method 0010* determines semi-volatile emissions.

(10) *Boilers and Industrial Furnaces (BIF) Methodology* determines metals emissions in exhaust gases from hazardous waste incineration and other combustion processes.

Pollutants that do not have established EPA test methods must be tested by other procedures developed by research laboratories, organizations and agencies.

10.4.3 CONTINUOUS EMISSION MONITORING (CEM)

The 1990 Clean Air Act Amendments and the associated state agency programs usually require CEM during the life of the facility if there are "significant" emissions (these are specified in the permit). Continuous monitoring regulations may require monitoring of:

- opacity
- flow
- temperature

- pressure and pressure differentials
- CO
- CO_2
- O_2
- SO_2
- other acid gases, e.g. HCl
- NO_x
- organics
- air toxics

An abbreviated summary of some characteristics of continuous emission monitors is given in Table 10.3. The US EPA has regulations to specify CEM performance criteria.

Some of the basic requirements are:

(1) *Relative accuracy (RA)*. A comparison of reference method values to actual CEM outputs. The CEM analyzer must produce results within

TABLE 10.3. **Summary of Some Continuous Emission Monitors.**

Pollutant	Monitor Type	Expected Concentration Range	Available Range[a]
O_2	Paramagnetic Electrocatalytic (e.g., zirconium oxide)	5–14%	0–25%
CO_2	NDIR[b]	2–12%	0–21%
CO	NDIR	0–100 ppm	0–5000 ppm
NO_x	Chemiluminescent	0–4000 ppm	0–10000 ppm
SO_2	Flame photometry Pulsed fluorescence NDUV[c]	0–4000 ppm	0–5000 ppm
SO_3	Colorimetric	0–100 ppm	0–50 ppm
Organic compounds	Gas chromatography (FID)[d] Gas chromatography (ECD)[e] Gas chromatography (PID)[f] IR absorption UV absorption GC/MS	0–50 ppm	0–100 ppm

[a]For available instruments only. Higher ranges are possible through dilution.
[b]Nondispersion infrared.
[c]Nondispersion ultraviolet.
[d]Flame ionization detector.
[e]Electron capture detector.
[f]Photo-ionization detector.

±20 percent of specified reference methods in a full set of test runs (9 to 12 runs of 30 minutes each). When measuring constituents in low concentrations (such as less than 30 parts per million), achieving this kind of agreement between an analyzer and reference method is difficult.

(2) *Calibration drift*. This is a measure of a unit's consistency over time, compared to National Bureau of Standards traceable calibration standards. Regulations require that drift over one day must not exceed four times the Performance Specification Test (PST) drift requirement, or the unit is considered "out of control." A unit is also "out of control" if the 5-day drift exceeds twice the PST drift. A CEM system that is out of control cannot be used for demonstrating compliance. When this occurs, it increases the chances that the unit will not meet its availability requirement.

(3) *Availability*. CEM systems must be available 90 percent of the boiler operating hours. Some states have stricter standards, such as higher percent availability and/or maintaining this availability within short operating periods. It is substantially more difficult, for example, to have data availability for 54 out of every 60 minutes (90 percent) than to have availability 22 out of every 24 hours (also 90 percent). Such requirements typically require facilities to install backup monitors for each pollutant and to operate both routinely, staggering calibration and maintenance schedules.

(4) *Data capture*. Instruments must achieve a minimum 75 percent valid data capture for each averaging period.

In addition to the basic continuous emission monitor itself, the CEM system is composed of three distinct subsystems: (1) the sampling interface, (2) the analytical instrumentation, and (3) the data acquisition system (DAS). The DAS typically has several data calculation and management, reporting, and archiving capabilities. It calculates the emission averages and outputs them to a variety of units, such as strip chart recorders. The sampling interface brings the flue gas into position for analysis. The type of sampling interface used characterizes the CEM technology. There are two major types: extractive and in-situ.

Extractive CEM systems extract samples from the flue gas, transport the sample, condition them for analysis, and record the sampled data. The gas sample is conditioned by filtering out particulate matter and removing water vapor or diluting the sample to a constant water content. Extractive systems require less maintenance than in-situ monitors and can be serviced by plant personnel with little specialized training. An additional advantage is that extractive monitors are specific for each pollutant measured. This increases the sensitivity for low emission pollutants and reduces errors that can be caused by interfering gases. However, extractive systems require an extensive on-site spare parts inventory, can be prone to plugging, and may have sluggish response times.

In-situ monitors analyze the flue gases "in place," without requiring sample transport or conditioning. In-situ monitors use light spectroscopy technology to measure components in flue gas—either at a single point or along a path of specific length in the duct or stack. Light passes from a transmitter-receiver through the gas to a reflector and back to the receiver. The reflected light is separated into differing wavelengths for measurement. A point analyzer uses electrochemical or electro-optical technology and is counted on the end of a probe at a single point in the flue gas. Path analyzers use electro-optical technology to measure gas concentrations along the entire path across the duct or stack.

In-situ monitors have quicker response time than extractive systems and require no sample conditioning. The gas being analyzed is not touched or affected by the analyzer. Disadvantages include high maintenance time and expense to maintain the complex optical and electronic systems, requirement for skilled service, and path length-dependent concentration ranges (sensitivity) for measurements.

Another disadvantage of in-situ systems is that they monitor all pollutants in a single unit. This is a drawback because, if the analyzer malfunctions, analysis for all measured pollutants ceases. This affects the system availability, making it difficult to meet permit standards (e.g., 90 percent data availability).

Examples of CEM systems and component typical costs in 1990 year dollars are listed in Table 10.4. Note that some items (e.g., the probe system) are shown with several options, and only one is required per CEM system. For an extractive CEM system, initially the sample gas is extracted from the stack gas by a sample probe located on the stack. The sample is then "cleaned" by the sample conditioning unit, which is normally mounted on the smoke stack as well. Thus, the sample first flows through the sample probe to a sample conditioning unit. Then the sample is sent through the stainless steel sample line to the analyzer located in the customer's control room. The calibration gas cylinders are used to calibrate the analyzers periodically to meet EPA requirements.

An extractive CEM system to comprise an EPA gas analysis system to monitor oxygen, carbon monoxide, carbon dioxide, and nitrogen oxides would normally be made up of the following components:

(1) Stainless steel sample probe mounted on the stack
(2) Sample conditioning unit mounted on the stack
(3) Sample line connecting the stack sample conditioning unit to the gas analyzer enclosures

TABLE 10.4. **Examples of Typical Extractive Systems and Components in 1990 Year Dollars.**

Description	Purchase Price, $
Portable probe system—50 cm. probe length—complete with ½% 30″ per hour servorecorder, air purge and automatic stack exit correction.	14,200
Same as above—100 cm probe length.	15,500
Opacity system—includes transmitter, receiver, and control unit. 50 ft maximum span, severe environment rated.	3,950
Opacity system—same as above, automatic zero and span system, 5′ maximum span. EPA compliance.	6,700
Baghouse monitor—transmitter/receiver, retroreflector, and control unit, 5 ft span.	1,975
Multipoint baghouse system, double pass system—adjustable set point with DPDT, 2A, relay contacts. External meter optional.	1,170/pt
Oxygen system, zirconia, in-situ—control unit, O_2—1V and 4–20 mA outputs. Meter scale .25 to 25% log output two decades. 2′ stainless steel probe.	2,650
Oxygen/combustible system, extractive—linear switchable scales, self diagnostic and relay contacts.	5,950
Combustibles system, extractive—linear switchable scales.	3,300
Combustibles system, in-situ—control unit, O–1V and 4–20 MA outputs. Switchable scales. 2′S/S probe.	3,250
SO_2—NDIR—EPA—sample conditioner, S/S probe, fiber glass case.	25,500
NO/NO_x, chemiluminescent method—auto span system with sample conditioner, EPA compliance with NEMA 12 enclosure.	30,200
Opacity system, single pass—dual beam for spans up to 50′. Contains transmitter, receiver, control box and lamp power supply.	12,950
Opacity system, double pass—dual beam, for spans to 185′. Contains transmitter, receiver, control box and lamp power supply. EPA compliance.	10,250

TABLE 10.5. Extractive CEM System to Monitor O_2, CO, CO_2 and NO_X in 1990 Year Dollars.

Description	Purchace Price, $
Less than 2′, ½″ stainless steel sample probe.	175
Sample conditioning unit containing the following: a. Heated filter for particulate removal b. Air operated chiller for water removal c. Coalescing filter d. Permapure dryer for additional water removal to keep the dew point in the −20 to −40°F region. System requires 4–6 CFM of plant air and a maximum 4 CFH of instrument air. Probe is back purged at predetermined intervals. Housed in a fiber glass, NEMA 4 enclosure.	included in below analyzers
Electric heat traced line. 3/8″ sample gas line. ¼″ cal. gas line. Insulated and jacketed weatherproof line, self-regulating includes signal cable between the sample conditioning unit and analyzer cabinet.	25/ft
Oxygen gas analysis system. This is an extractive system using a heated zirconia sensor. Contains coalescing inlet filter and vacuum pump.	3,200
Carbon monoxide gas analysis system meeting EPA requirements. The carbon monozide (CO) component in the sample gas is analyzed using the non-dispersive infrared (NDIR) technique. This method uses two cells, a sample and reference cell. A dual beam IR light source is projected through each cell and this energy is alternately shifted to a detector cell by a light chopper which operates at about 10 revolutions/second. The detector cell is the Luft type which acts as a mechanical variable capacitor. The cell is filled with the same CO concentration as the reference cell. The capacitance of the detector cell is a function of the CO concentration. The cell is part of LC (inductance, capacitance) resonant tank circuit which is connected to an oscillator. The frequency of oscillation is now a function of the concentration of CO gas in the sample stream. A frequency to voltage circuit converts the signal to a DC voltage.	25,500
Carbon dioxide gas analysis system meeting EPA requirements. The CO_2 component of the gas stream is measured using a similar NDIR method as described for the CO system.	included in above CO price
The analyzer, pumps, automatic timing system for EPA calibration and all other controls and gauges are housed in a standup enclosure.	included in above analyzers
In the EPA system, the solenoid valves that control the calibration gases may be in the standup enclosure or at the sample probe.	included in above analyzers

TABLE 10.5. (continued).

Description	Purchace Price, $
NO/NO_x gas analysis system meeting EPA requirements. This is an extractive analyzer using the chemiluminescent method. The stack gas sample is conditioned or "cleaned" of water and particulate matter by the sample conditioning unit before the sample is inserted into the analyzer. There are two methods of analysis. No_x consists of nitrogen oxide and nitrogen dioxide. To measure NO_x the sample gas flows through a temperature chamber (converter) that is heated to about 600°F. The nitrogen dioxide component is converted to NO, so that only NO flows into the combustion chamber.	
Ozone (O_3) is added to the NO in the combustion chamber and a photomultiplier tube (PMT) measures the small amount of light emitted by the reaction of NO + O_3.	
To measure NO, the converter is bypassed and the sample gas stream goes straight into the combustion chamber where it is mixed with the ozone. Again, the PMT measures the light emitted by the reaction. The nitrogen dioxide component is the difference between the NO_x and NO.	
To measure NO, the converter is bypassed and the sample gas stream goes straight into the combustion chamber where it is mixed with the ozone. Again, the PMT measures the light emitted by the reaction. The nitrogen dioxide component is the difference between the NO_x and NO.	
The intensity of the emitted light which is in the light wavelength region of 590 – 2705 nm is proportional to the mass flow rate of NO in the combustion chamber. The voltage from the PMT is converted to a low impedance output voltage signal and a 4 – 20 mA interface.	
The PMT is controlled at 6°C by a thermoelectric cooler for stable operation.	
The ozone generator, which is internal to the instrument, uses an ultraviolet technique to generate ozone from an instrument air source.	28,050
This system does not include calibration gas cylinders or regulators.	Not included
4″ felt tip, z fold, 3 pen recorder	3,475

(4) Gas analyzers, pumps, etc., housed in a standup enclosure for O_2, CO, CO_2, NO_x.

(5) Solenoid valves that control the calibration gases may be in the standup enclosure or at the sample probe.

Table 10.5 gives detailed descriptions and purchase prices for an EPA quality system to enable a facility to set up a system to carry out this monitoring. Installed costs would normally be about twice the total purchase cost.

List of Abbreviations

AACL – acceptable ambient concentration levels
A/C – air to cloth ratio
ACFM – actual cubic feet per minute
APC – best available control technology
BIF – boiler and industrial furnace technology
CAA – clean air act
CAAA – Clean Air Act Amendments
CEM – continuous emission monitor
CFR – Code of Federal Regulations
DSCF – dry standard cubic foot
DSI – dry solvent injection
EPA – Environmental Protection Agency
ESP – electrostatic precipitation
FGD – flue gas desulfurization
g – grams
GPM – gallons per minute
gr – grains (1/7000 of a pound)
HAPs – hazardous air pollutants
LAER – lowest achievable emission rate
MACT – maximum achievable control technology
MSDS – material safety data sheets
NAAQS – national ambient air quality standards
NESHAP – national emission standards for hazardous air pollutants
NSPS – new source performance standards
OSHA – Occupational Safety and Health Administration
OVA – organic vapor analyzer

PICs – products of incomplete combustion
ppm – parts per million (by volume for gases and by mass for liquids and solids)
ppmv – parts per million by volume
PSD – prevention of significant deterioration
RACT – reasonably available control technology
RCRA – Resource Control and Recovery Act
scf – standard cubic feet
SIP – state implementation plan
TLV – threshold limit value
VOCs – volatile organic compounds
w.c. – water column (pressure drops)

Index